Analysis of the Alkali Metal Diatomic Spectra

Diatomic Spectra

Using molecular beams and ultracold molecules

Analysis of the Alkali Metal Diatomic Spectra

Using molecular beams and ultracold molecules

Jin-Tae Kim

Department of Photonic Engineering, Chosun University, Gwangju, Korea

Bongsoo Kim

Department of Chemistry, KAIST, Daejeon, Korea

William C Stwalley

Department of Physics, University of Connecticut, Storrs, USA
Also, *Kavli Institute for Theoretical Physics, University of California at Santa Barbara, Santa Barbara, USA (January–March 2013)*

Morgan & Claypool Publishers

Rights & Permissions
To obtain permission to re-use copyrighted material from Morgan & Claypool Publishers, please contact info@morganclaypool.com.

ISBN   978-1-6270-5678-6 (ebook)
ISBN   978-1-6270-5677-9 (print)
ISBN   978-1-6270-5679-3 (mobi)

DOI   10.1088/978-1-6270-5678-6

Version: 20141201

IOP Concise Physics
ISSN 2053-2571 (online)
ISSN 2054-7307 (print)

A Morgan & Claypool publication as part of IOP Concise Physics
Published by Morgan & Claypool Publishers, 40 Oak Drive, San Rafael, CA, 94903, USA

IOP Publishing, Temple Circus, Temple Way, Bristol BS1 6HG, UK

Contents

Preface

This ebook illustrates the complementarity of molecular beam (MB) spectra and ultracold molecule (UM) spectra in unraveling the complex electronic spectra of diatomic alkali metal molecules, using KRb as a prime example. Researchers interested in molecular spectroscopy, whether physicist, chemist, or engineer, may find this ebook helpful and may be able to apply similar ideas to their molecules of interest.

Diatomic molecules are the simplest of molecules and the two alkali metal atoms which make up these particular diatomic molecules are the simplest of atoms, except for hydrogen atoms. Hence theoretical understanding of the molecules presented here is comparable to the understanding of H_2 and the alkali metal hydrides, e.g. LiH. However, even these simple molecules pose challenges to the spectroscopist. Two special techniques, molecular beam (MB) and ultracold molecule (UM) spectroscopy described herein, provide special information that can be used to overcome these challenges.

One challenge is that observable concentrations of alkali metal diatomic molecules are normally produced at high temperatures. However, at high temperatures dozens of vibrational states and hundreds of rotational states are typically produced, so that assigning the spectra is difficult without extensive use of multi-laser, multiple-resonance spectroscopy. In contrast, in MB spectroscopy (with typical vibrational and rotational temperatures of a few K), a large majority of the population (typically > 95%) is in the lowest vibrational state ($v'' = 0$) of the $X\,^1\sum^+$ ground singlet electronic state and the most populated rotational level is typically $J'' = 5$. In UM spectroscopy, the rotational temperature is typically even colder, but the molecules are in very high vibrational levels near the dissociation limit of two ground state atoms in the $X\,^1\sum^+$ ground singlet electronic state and/or the $a\,^3\sum^+$ metastable triplet electronic state. Thus the states studied by MB spectroscopy are those centered around the equilibrium distance of the X state, R_e, while the states studied by UM spectroscopy are those centered about very large internuclear distances, $R \gg R_e$, where the interaction between the two alkali atoms becomes very weak.

A second challenge is that the molecules in a high temperature oven are virtually all singlet molecules, which is also true in a MB. However, since UM spectra produce singlet X state and/or triplet a state molecules, spectroscopy among triplet electronic states can easily be studied. Moreover, the so-called perturbed levels, which are neither singlet nor triplet but a mixture of the two, can be identified and can be used as 'windows' to transfer molecules from singlet to triplet or from triplet to singlet with ease.

In this ebook, after an introduction (chapter 1), we describe in detail supersonic MB experiments, emphasizing the pulsed techniques, and then UM formation based on photoassociation (PA) (chapter 2). The conventional spectroscopic results for the highlighted molecule, KRb, are then described, followed by discussions of the MB and UM spectroscopy of KRb and the corresponding improvements, and finally by discussions of the prospects for other heteronuclear and homonuclear alkali metal diatomic molecules (chapter 3). We then discuss multiplication (product) spectroscopy, MBxUM, which is a powerful new kind of spectrum, particularly useful in optimizing Raman transfer routes, even when the individual MB and UM spectra have not been fully assigned (chapter 4). This is followed by a brief conclusion (chapter 5).

Acknowledgements

This research was supported in the USA by the NSF and the AFOSR (Multi-university Research Initiative on Ultracold Polar Molecules). In addition, WCS was partially supported during the winter of 2013 by the Kavli Institute for Theoretical Physics at the University of California at Santa Barbara. This research was supported at Chosun University (2012-0085319) and at KAIST (2011-0001335 and 2011-0020419) by NRF in Korea.

The authors

Jin-Tae Kim

Jin-Tae Kim is a Professor of the Department of Photonic Engineering at Chosun University, Gwangju, Korea. He has served as the chair of the division of atomic and molecular physics in the Korean Physical Society since 2014. He received BS and MS degrees in Physics from Korea University, Seoul, Korea, in 1984 and 1986, and a PhD degree from the University of Iowa, Iowa City, USA in 1995. His dissertation work involved all-optical multiple resonance spectroscopy of the potassium diatomic molecule using a high resolution CW ring dye and Ti:Sapphire lasers. In 1995, he joined the Physics Department, University of Connecticut, Storrs, where he performed laser molecular supersonic beam experiments for hydrogen molecules. In 1997, he joined the laboratory for Quantum Optics, Korea Atomic Energy Research Institute, where he was engaged in the research and development of dye lasers and laser atomic spectroscopy by using a time of mass spectrometer. His primary research interest is laser ultracold atomic and molecular spectroscopy. Other research interests include laser applications such as display and information storage using digital holography, pattern recognition using optical correlation methods, surface profile measurements using the interferometer, and ultra-sensitive detection of gas.

Bongsoo Kim

Bongsoo Kim is Professor of Chemistry at KAIST in Korea. He has a BS and a MS degree in Chemistry from Seoul National University and a PhD in Physical Chemistry from U. C. Berkeley. He spent two years at Kyungbook National University as a professor. Then he moved to KAIST in 1996, serving there at present. His primary interest has been laser spectroscopy in molecular beams. Recently he expanded his interest into nanoscience and synthesized gold nanowires and nanoplates, utilizing them for surface enhanced Raman scattering. In 2015 he serves as Chairman of the Physical Chemistry Section of the Korean Chemical Society for a year. Major awards include the Grand Research Prize in 2011 from the Korean Chemical Society.

William C Stwalley

William C Stwalley is Board of Trustees Distinguished Professor of Physics at the University of Connecticut at Storrs as well as Affiliate Professor of Chemistry and member of the Institute of Materials Science. He has a BS degree in Chemistry from Caltech and a PhD in Physical Chemistry from Harvard. He next spent 25 years at the University of Iowa, where he rose through the ranks to the George Glockler Professor of Physics and Chemistry. In 1993, he became Department Head of the Physics Department at UConn, serving in that capacity until 2011. His primary interest has been atomic and molecular interactions, which he has studied theoretically and experimentally throughout his career, emphasizing the determination of potential energy curves out to long range and the use of laser spectroscopy. In addition to fellowship in the American Association for Advancement of Science, the American Physical Society and the Optical Society of America, major awards received include the Meggers Award for Spectroscopy of the Optical Society of America and the Connecticut Medal of Science.

IOP Concise Physics

Analysis of the Alkali Metal Diatomic Spectra
Using molecular beams and ultracold molecules

Jin-Tae Kim, Bongsoo Kim and William C Stwalley

Chapter 1

Introduction

Since 1911, many experimental and theoretical methods [1–16] for reducing the velocities and internal energies of molecules have been proposed and implemented. The slowing and cooling down of molecules has provided new fascinating and challenging opportunities such as new reaction dynamics [17–24], chemical reaction control [25–29], formation of dense and stable ensembles of molecules [14, 30–34], and precision measurements [35, 36] of physical quantities, such as the electric dipole moment of the electron.

Molecules, with rotational, vibrational, and electronic internal structure, and often a permanent electric dipole moment, are complex compared to atoms with only electronic structure and no dipole moment [10, 37]. It is not easy to control and manipulate the many degrees of freedom of molecules. However, molecules give a larger set of information to the scientists if those degrees of freedom can be controlled, for example, by interaction with external physical parameters [37] such as electric and magnetic fields. To manipulate the molecules with external fields, it is necessary to address more constraints for the preparation of molecules in a single quantum state, and often to increase the interaction time between molecules and external fields.

In the mid 20th century, an effusive MB apparatus [38, 39], modified from an atomic beam [40, 41], was designed to select a specific component of molecular velocity. To achieve additional reduction of molecular velocities, a seeded supersonic MB method with skimmer [2] was developed to reduce the internal energy of molecules more than the effusive MB apparatus although this seeded method accelerates molecules, e.g. using He, Ne, etc. This method provides molecules with kinetic energy within a narrow region and with very low internal temperature. The Stark decelerator [3], in combination with a supersonic MB was also employed to further reduce the velocities of the molecules.

Although alkali metals with high evaporation temperatures and high reactivity are difficult to control experimentally, investigations into spectroscopic assignments

doi:10.1088/978-1-6270-5678-6ch1

for rovibrational levels of electronic states of diatomic alkali metal molecules in supersonic MBs have been studied by many groups [42–127]. In particular, pulsed supersonic MBs [91, 97, 105–111, 119–127] achieve much colder internal temperatures than CW supersonic MBs.

Other ways of cooling down molecules include cryogenic buffer gas cooled beams [4, 6], and also indirect and direct optical cooling of the molecules [8, 12, 128]. Recent optical cooling and trapping of the YO molecule [7] also achieved the order of mK temperature. Up to now, the typical temperature obtained by the buffer gas cooling of molecules is several mK, approaching the temperatures obtained by optical methods. However, the temperature is far above the temperature of nK obtained in atoms, where Bose–Einstein condensates [129, 130] were obtained in 1995.

There are two other techniques, photoassociation (PA) and magnetoassociation (MA), which, unlike the techniques discussed above, form cold diatomic alkali molecules from ultracold alkali atoms. Photoassociated molecules can be formed by illuminating ultracold atoms with a PA laser. The photoassociated molecules typically decay to form cold molecules in high vibrational levels of the $X\,^1\Sigma^+$ and $a\,^3\Sigma^+$ states instead of decaying to lower vibrational levels of the $X\,^1\Sigma^+$ and $a\,^3\Sigma^+$ states. There have been many spectroscopic investigations [131–202] of photoassociated and ultracold molecules (UM).

The MA method, by ramping a magnetic field, forms diatomic molecules in very high energy levels near the dissociation limit of the ground state, ultimately including quantum degenerate molecular ensembles of so-called Feshbach molecules [203–205]. The near dissociation levels of UMs can be formed most efficiently at nK temperatures by forming Feshbach molecules with predominantly triplet character through MA from ultracold atoms.

However, the photoassociated and Feshbach molecules are in very high vibrational and low rotational levels of the $X\,^1\Sigma^+$ and $a\,^3\Sigma^+$ states near the dissociation limit of two ground state alkali atoms. Even heteronuclear alkali diatomic molecules in high vibrational levels have a very small electric dipole moment so that it is necessary to transfer the population to a much lower level of the ground state to enhance the dipole. In addition, molecules in the lowest rovibrational levels of the X and a states (denoted $X(0,0)$ and $a(0,0)$, respectively) cannot undergo rovibrationally inelastic collisions at ultracold temperatures. Thus it is often important to transfer the population of highly excited molecular levels (near the dissociation limit of two ground state atoms) to $X(0,0)$ and $a(0,0)$ through intermediate electronic states. There are two ways to generate $X(0,0)$ and $a(0,0)$ molecules, through coherent stimulated Raman adiabatic passage (STIRAP) and through incoherent decay.

We focus here on heteronuclear diatomic molecular cases (especially KRb) for transfer to the lowest level of the $X(0,0)$ ground state using MA and PA coherently or incoherently. Magnetoassociated $^{40}\mathrm{K}^{87}\mathrm{Rb}$ molecules [19, 25] have been formed efficiently in near dissociation levels and transferred to $X(0,0)$ using STIRAP through mixed singlet–triplet $1\,^1\Pi \sim 2\,^3\Sigma^+$ levels. These molecules have also had their population subsequently transferred to the lowest hyperfine state [25]. Finally these molecules have been used to manipulate ultracold chemical reactions by controlling quantum states [25, 26]. Aikawa *et al* [206] formed ultracold $^{41}\mathrm{K}^{87}\mathrm{Rb}$ molecules in the lowest level ($v'' = 0$, $J'' = 0$) of the ground $X\,^1\Sigma^+$ state through PA followed by an efficient STIRAP transfer process. This STIRAP transfer to the lowest rovibrational level was the first coherent transfer achievement from photoassociated molecules in a high vibrational level ($v'' = 91$, $J'' = 0$ and 2) of the X state made by using an intermediate level ($v' = 41$, $J' = 1$) of the singlet $3\,^1\Sigma^+$ state, as proposed by Wang *et al* [178]. The narrow linewidth and relatively weak transition dipole moment (TDM) [207] between high vibrational levels of the X state and vibrational levels of the $3\,^1\Sigma^+$ state should be noted.

Incoherent transfer produces molecules distributed over a broad range of vibrational levels of the ground state. The molecules in levels with residual internal energies then undergo inelastic collisions which heat the sample. The lowest level ($v'' = 0$) of the ground $X\,^1\Sigma^+$ state was formed via a vibrational level mixed among the $2\,^3\Sigma^+$, $1\,^1\Pi$, and $b\,^3\Pi$ states from a specific high vibrational level ($a\,^3\Sigma^+$ ($v_a = 37$)) near dissociation (formed from photoassociated RbCs molecules [196]). The RbCs molecule formed in the lowest level ($v'' = 0$) of the ground $X\,^1\Sigma^+$ state had a translational temperature of ~100 μK. PA to a rovibrational level ($v' = 4$, $J' = 1$) of the $1\,^1\Pi$ state at shorter internuclear distance from trapped Li and Cs atoms using a forced dark-spot magneto-optical trap (MOT) led to decay into the lowest rovibrational level of the ground state directly from the upper PA level [173]. Zabawa *et al* [188, 191] produced ultracold NaCs molecules in the lowest vibrational level ($v'' = 0$, $J'' = 1$) of the $X\,^1\Sigma^+$ state incoherently from two PA levels ($2\,(0^+)$, $J' = 1$ and 2) coupled with the short range electronic $2\,^1\Pi$ state. Recently, ultracold ^{39}K^{85}Rb molecules in the lowest vibrational levels ($v'' = 0$–10) were incoherently formed from a resonantly coupled PA level between the $2(1)$, $v' = 165$ and $4(1)$, $v' = 61$ levels [208].

These transfer studies should significantly advance the perspectives for the production, manipulation, and application of polar ultracold molecules, the mutual interaction of which is governed by highly anisotropic electric dipole–dipole forces promising qualitatively new dynamical regimes unavailable in ultracold homonuclear systems. Such strongly interacting molecules have been suggested for use as the qubits [37] of a scalable quantum computer. The control of ultracold chemical reactions has been proposed by using weakly bound, electric field-linked states of participating molecules. The large polarizability of polar molecules strongly enhances parity-violation effects caused by the electric dipole moment of the electron. Cooling such molecules down to ultracold temperatures should dramatically increase coherence times and, therefore, the sensitivity in measurements of these effects.

In order to control and manipulate heteronuclear alkali dimers, highly accurate spectroscopic data on electronically excited states, as well as the ground state, are indispensable. However, it is often difficult to understand all spectroscopic information of simple diatomic molecules which have global perturbations and abnormal potential wells. The available data is often limited to a few electronic states. Although high-resolution Fourier transform spectroscopy [209–210] has provided very accurate spectroscopic data for many of the alkali dimers, it is limited to low lying electronic states. RE2PI (resonance enhanced two-photon ionization) spectroscopy using MBs has been successful in the detection of very weak transitions [91, 119, 123] and assignments of complex spectra [121, 122, 126, 127], but it is still not easy to assign all of the severely perturbed spectroscopic lines [181, 182] and all of the spin–orbit component assignments. Also, ultracold molecule (UM) spectroscopy cannot easily determine all those assignments.

Here we have been combining the results of MB and UM spectroscopic experiments on KRb [181, 182]. In particular, the combination allows definitive assignments of nearly all lines in both spectra if the energy regions of the excited states involved are the same. Moreover, multiplicative spectra between the MB and UM spectra [211, 212] can be used to find the optimum pathways for stimulated Raman population transfer from translationally UMs, typically formed in high rovibrational levels near the ground state dissociation limit, to fully translationally and rovibrationally UMs in the lowest rovibrational level of the ground electronic state. The maximum intensity peaks of these product spectra indicate the optimal pathways for the population transfer, which can be obtained without spectral assignment. We illustrate this for KRb (which is frequently used for applications of UMs).

This multiplication intensity, $I(SR)$, between the MB and UM spectra for SR population transfer from a specific high vibrational level (v, J) with a low rotational quantum number of the $a\,^3\Sigma^+$ or $X\,^1\Sigma^+$ states to the $(v'' = 0, J'' = 0)$ levels of the $X\,^1\Sigma^+$ state via an $|e', v', J'\rangle$ intermediate level can be described as

$$I(SR) \propto \left| \langle a \text{ or } X, v, J | d_{a \text{ or } X,e'} | e', v', J' \rangle \right|^2 \times \left| \langle e', v', J' | d_{e',X} | X, v'' = 0, J'' = 0 \rangle \right|^2$$

(1.1)

where $d_{a \text{ or } X,e'}$ and $d_{e',X}$ are the corresponding electric dipole transition moment functions. In this equation the first matrix element squared is proportional to the transition intensity from the initial $|a\,^3\Sigma^+ \text{ or } X\,^1\Sigma^+, v, J\rangle$ UM level to the $|e', v', J'\rangle$ intermediate level, proportional to the intensity of the UM spectra. The second matrix element squared is proportional to the transition intensity from the $|X\,^1\Sigma^+, v'' = 0, J''\rangle$ MB level to the same $|e', v', J'\rangle$ intermediate level and is proportional to the intensity of the MB spectra. Thus, the relative efficiency of various SR pathways can be obtained by simply taking the product of the two transition intensities observed in the UM and MB spectra. This approach determines the laser frequencies required for optimal SR transfers without the need to identify the intermediate $|e', v', J'\rangle$ levels. Thus the maximum value of this result (the product MB × UM at each upper energy level) provides the optimal SR pathways without spectral analysis of the excited electronic states.

Figures 1.1 and 1.2 show STIRAP population transfer schemes to the lowest energy level from UMs by using an intermediate singlet state (sSR) as well as states of mixed singlet and triplet character (mSR) to obtain the intensities of the multiplication spectra, respectively. The relevant *ab initio* potential energy curves (PECs) [213] of the ground and excited electronic states below and above the first excited asymptotic dissociation limit, primary energy levels, and PA and SE involved in the case of KRb are also shown in figure 1.1. Here, MB and UM spectra are used for mSR2 (or sSR2) and mSR1 (or sSR1) processes, respectively.

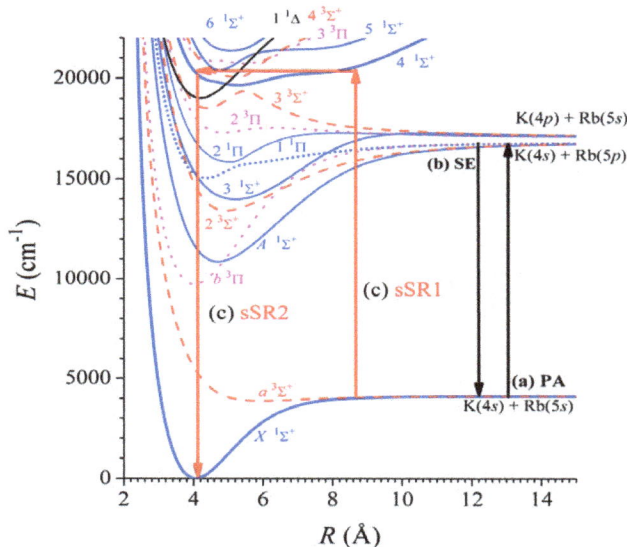

Figure 1.1. Experimental Raman transfer scheme for the multiplication between MB and UM spectra, the relevant *ab initio* PECs [213], primary energy levels, and processes involved in this experiment: (*a*) PA to specific rovibrational levels of the $3(0^+)$ state from two colliding ultracold K and Rb atoms to a bound electronically excited state with rovibrational quantum numbers v', J'; (*b*) Subsequent spontaneous emission (SE) to the weakly bound $X\,^1\Sigma^+(v'', J'')$ levels; (*c*) SRT combines the MB and UM transitions: sSR1 and sSR2 take population from the weakly-bound $X\,^1\Sigma^+(v'', J'')$ levels to the $X\,^1\Sigma^+(v''=0, J''=0)$ level via the pure singlet $e'(v', J')$ intermediate level. Their resonant absorption scheme from the UM in the weakly bound X $^1\Sigma^+(v'', J'')$ levels to vibrational levels of the excited intermediate states, followed by ionization to $KRb^+ + e^-$ is not shown. Also, the resonant absorption scheme from the $X\,^1\Sigma^+$ $(v''=0, J''=0)$ level to vibrational levels of the excited intermediate states, followed by ionization to $KRb^+ + e^-$ (REMPI) is not shown.

In this ebook, the population transfer of alkali metal diatomic molecules to the lowest rovibrational level of the ground state from molecules formed by PA and MA has been reviewed in chapter 1.

In chapter 2 we review how **MB** experiments have been developed and applied to alkali metal diatomic molecules (section 2.1), especially in the case of KRb (see also section 3.2); we describe our specific MB experiments (section 2.2); and we describe our specific UM experiments (section 2.3).

In chapter 3 we review the experimental spectroscopy of KRb in a heat-pipe oven (section 3.1), in a MB apparatus (section 3.2), and in an UM apparatus (section 3.3), and demonstrate the utility of analyzing mutually perturbing states using the multiplication spectrum denoted as MB × UM (section 3.4). We also discuss the prospects for similar studies for other alkali metal diatomics beyond KRb (section 3.5).

In chapter 4 we show how the multiplication spectrum (MB × UM) can also be used to determine optimal Raman transfer paths in KRb, even in the absence of detailed analysis of the mutually perturbing states and finally we draw conclusions in chapter 5.

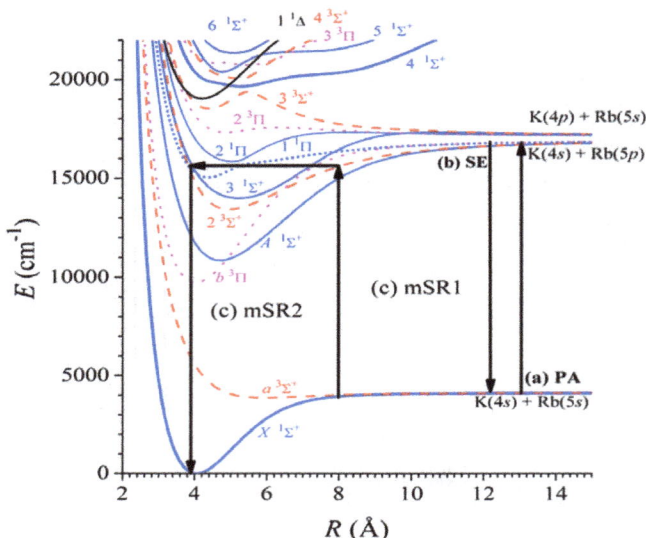

Figure 1.2. Experimental Raman transfer scheme for the multiplication between MB and UM spectra, the relevant *ab initio* PECs [213], primary energy levels, and processes involved in this experiment: (*a*) PA to specific rovibrational levels of the 3(0$^+$) or 3(0$^-$) state from two colliding ultracold K and Rb atoms to a bound electronically excited state with rovibrational quantum numbers v', J'; (*b*) Subsequent spontaneous emission (SE) to the weakly bound $a\,^3\Sigma^+(v_a, J_a)$ or $X\,^1\Sigma^+(v'', J'')$ levels; (*c*) SRT combines the MB and UM transitions: mSR1 and mSR2 take population from the weakly-bound $a\,^3\Sigma^+(v, J)$ or $X\,^1\Sigma^+(v'', J'')$ levels to the $X\,^1\Sigma^+(v'' = 0, J'' = 0)$ level via the singlet-and-triplet-mixed $e'(v', J')$ intermediate level (or the singlet $e'(v', J')$ intermediate level). Their resonant absorption scheme from the UM in the weakly bound $a\,^3\Sigma^+(v_a, J_a)$ or $X\,^1\Sigma^+(v'', J'')$ levels to vibrational levels of the excited intermediate states, followed by ionization to $KRb^+ + e^-$ is not shown. Also, the resonant absorption scheme from the $X\,^1\Sigma^+(v'' = 0, J'' = 0)$ level to vibrational levels of the excited intermediate states, followed by ionization to $KRb^+ + e^-$ (REMPI) is not shown.

References

[1] Dunoyer L 1911 Réalisation d'un rayonnement matériel d'origine purement thermique *J. Radium* **8** 142

[2] Smalley R E, Ramakrishna B L, Levy D H and Wharton L 1974 Laser spectroscopy of supersonic molecular beams: application to the NO$_2$ spectrum *J. Chem. Phys.* **61** 4363

[3] Bethlem H L, Berden G, Crompvoets F M N, Jongma R T, van Roij A J A and Meijer G 2000 Electrostatic trapping of ammonia molecules *Nature* **406** 491

[4] Hutzler N R, Lu H-I and Doyle J M 2012 The buffer gas beam: an intense, cold, and slow source for atoms *Chem. Rev.* **112** 4803

[5] Narevicus E and Raizen M 2012 Toward cold chemistry with magnetically decelerated supersonic beams *Chem. Rev.* **112** 4879

[6] Weinstein J D, de Carvalho R, Guillet T, Friedrich B and Doyle J M 1998 Magnetic trapping of calcium monohydride molecules at millikelvin temperatures *Nature* **395** 148

[7] Hummon M T, Benjamin M Y, Stuhl K, Collopy A L, Xia Y and Ye J 2013 2D magneto-optical trapping of diatomic molecules *Phys. Rev. Lett.* **110** 143001

[8] Schuman E S, Barry J F and DeMille D 2010 Laser cooling of a diatomic molecule *Nature* **467** 820

[9] Thorsheim H R and Weiner J 1987 Laser-induced photoassociation of ultracold sodium atoms *Phys. Rev. Lett.* **58** 2420

[10] Bahns J T, Stwalley W C and Gould P L 1996 Laser cooling of molecules: a sequential scheme for rotation, translation, and vibration *J. Chem. Phys.* **104** 9689

[11] Stwalley W C and Wang H 1999 Photoassociation of ultracold atoms: a new spectroscopic technique *J. Mol. Spectrosc.* **195** 194

[12] Fioretti A, Comparat D, Crubellier A, Dulieu O, Masnou-Seeuws F and Pillet P 1998 Formation of cold Cs_2 molecules through photoassociation *Phys. Rev. Lett.* **80** 4402

[13] Nikolov A N, Ensher J R, Eyler E E, Wang H, Stwalley W C and Gould P L 2000 Efficient production of ground-state potassium molecules at sub-mK temperatures by two-step photoassociation *Phys. Rev. Lett.* **84** 246

[14] Wynar R, Freeland R S, Han D J, Ryu C and Heinzen D J 2000 Molecules in a Bose-Einstein condensate *Science* **287** 1016

[15] van de Meerakker S Y T, Bethlem H L, Vanhaecke N and Meijer G 2012 Manipulation and control of molecular beams *Chem. Rev.* **112** 4828

[16] Stuhl B K, Sawyer B C, Wang D and Ye J 2008 Magneto-optical trap for polar molecules *Phys. Rev. Lett.* **101** 243002

[17] Que'me'ner G and Bohn J L 2011 Dynamics of ultracold molecules in confined geometry and electric field *Phys. Rev.* A **83** 012705

[18] Zirbel J J, Ni K-K, Ospelkaus S, D'Incao J P, Wieman C E, Ye J and Jin D S 2008 Collisional stability of Fermionic Feshbach molecules *Phys. Rev. Lett.* **100** 143201

[19] Ni K-K, Ospelkaus S, Wang D, Que'me'ner G, Neyenhuis B, de Miranda M H G, Bohn J L, Ye J and Jin D S 2010 Dipolar collisions of polar molecules in the quantum regime *Nature* **464** 1324

[20] Wang T T, Heo M-S, Rvachov T M, Cotta D A and Ketterle W 2013 Deviation from universality in collisions of ultracold $^6Li_2$ molecules *Phys. Rev. Lett.* **110** 173203

[21] Chin C, Kraemer T, Mark M, Herbig J, Waldburger P, Nägerl H-C and Grimm R 2005 Observation of Feshbach-like resonances in collisions between ultracold molecules *Phys. Rev. Lett.* **94** 123201

[22] Byrd J N, Montgomery J A and Côté R 2012 Controlling binding of polar molecules and metastability of one-dimensional gases with attractive dipole forces *Phys. Rev. Lett.* **109** 083003

[23] Chin C, Grimm R and Julienne P S 2010 Feshbach resonances in ultracold gases *Rev. Mod. Phys.* **82** 1225

[24] Carr L D, DeMille D, Krems R V and Ye J 2009 Cold and ultracold molecules: science, technology and applications *New J. Phys.* **11** 055049

[25] Ospelkaus S, Ni K-K, Que'me'ner G, Neyenhuis B, Wang D, deMiranda M H G, Bohn J L, Ye J and Jin D S 2010 Controlling the hyperfine state of rovibronic ground-state polar molecules *Phys. Rev. Lett.* **104** 030402

[26] Ospelkaus S, Ni K-K, Wang D, de Miranda M H G, Neyenhuis B, Que'me'ner G, Julienne P S, Bohn J L, Jin D S and Ye J 2010 Quantum-state controlled chemical reactions of ultracold potassium-rubidium molecules *Science* **327** 853

[27] de Miranda M H G, Chotia A, Neyenhuis B, Wang D, Ospelkaus S, Bohn J L, Ye J and Jin D S 2011 Controlling the quantum stereodynamics of ultracold bimolecular reactions *Nature Phys.* **7** 502

[28] Gaubatz U, Rudecki P, Becker M, Schiemann S, Külz M and Bergmann K 1988 Population switching between vibrational levels in molecular beams *Chem. Phys. Lett.* **149** 463

[29] Krems R V, Stwalley W C and Friedrich B 2009 *Cold molecules: theory, experiment, applications* (Boca Raton, FL: Taylor and Francis)

[30] Inouye S, Andrews M R, Stenger J, Miesner H-J, Stamper-Kurn D M and Ketterle W 1998 Observation of Feshbach resonances in a Bose-Einstein condensates *Nature* **392** 151

[31] Greiner M, Regal C A and Jin D S 2003 Emergence of a molecular Bose-Einstein condensate from a Fermi gas *Nature* **426** 537

[32] Jochim S, Bartenstein M, Altmeyer A, Hendel G, Riedel S, Chin C, Denschlag J H and Grimm R 2003 Bose-Einstein condensation of molecules *Science* **302** 2101

[33] Ni K-K, Ospelkaus S, de Miranda M H G, Pe'er A, Neyenhuis B, Zirbel J J, Kotochigova S, Julienne P S, Jin D S and Ye J 2008 A high phase-space-density gas of polar molecules *Science* **322** 231

[34] Bruun G M and Taylor E 2008 Quantum phases of a two-dimensional dipolar Fermi gas *Phys. Rev. Lett* **101** 245301

[35] DeMille D, Cahn S B, Murphree D, Rahmlow D A and Kozlov M G 2008 Using molecules to measure nuclear spin-dependent parity violation *Phys. Rev. Lett.* **100** 023003

[36] Hudson J J, Kara D M, Smallman I J, Tarbutt M R and Hinds E A 2011 Improved measurement of the shape of the electron *Nature* **473** 493

[37] DeMille D 2002 Quantum computation with trapped polar molecules *Phys. Rev. Lett.* **88** 067901

[38] Kistiakowsky G B and Slichter W P 1951 A high intensity source for the molecular beam. Part II. Experimental *Rev. Sci. Instrum.* **22** 333

[39] Ramsey N 1956 *Molecular beams* (Oxford University Press: Oxford)

[40] Zacharias J R 1942 The nuclear spin and magnetic moment of ^{40}K *Phys. Rev.* **61** 270

[41] Cohen V W and Ellett A 1937 Velocity analysis by means of the Stern-Gerlach effect *Phys. Rev.* **52** 502

[42] Foster P J, Leckenby R E and Robbins E J 1969 The ionization potentials of clustered alkali metal atoms *J. Phys.* B **2** 478

[43] Wu C-Y R, Crooks J B, Yang S C, Way K R and Stwalley W C 1973 Li/Li$_2$ supersonic nozzle beam *Rev. Sci. Instrum.* **49** 380

[44] Mathur B P, Rothe E W, Reck G P and Lightman A J 1981 Two-photon ionization of Li$_2$: isotopic separation and determination of IP(Li$_2$) and D_e(Li$^+_2$) *Chem. Phys. Lett.* **56** 336

[45] Eisel D and Demtröder W 1982 Accurate ionization potential of Li$_2$ from resonant two-photon ionization *Chem. Phys. Lett.* **88** 481

[46] Engelke F and Hage H 1983 Direct observation of the lowest b $^3\Pi_u$ state of the ^{6}Li$_2$ molecule *Chem. Phys. Lett.* **103** 98

[47] Rubahn H-G and Toennies J P 1988 A molecular beam study of the potential anisotropy of laser vibrationally excited Li$_2$ ($v = 0$, 20) scattered from Kr *J. Chem. Phys.* **89** 287

[48] Schwarz M, Duchowicz R, Demtröder W and Jungen Ch 1988 Autoionizing Rydberg states of Li$_2$: Analysis of electronic–rotational interactions *J. Chem. Phys.* **89** 5460

[49] Ishikawa K, Kubo S and Katô H 1991 The Li$_2$ C $^1\Pi_u$ state studied by a single-frequency ultraviolet laser *J. Chem. Phys.* **95** 8803

[50] Bouloufa N, Cacciani P, Vetter R and Yiannopoulou A 1999 Tunneling through potential barrier of the B $^1\Pi_u$ state of ^{7}Li-^{7}Li *J. Chem. Phys.* **111** 1926

[51] Bouloufa N, Cacciani P, Vetter R, Yiannopoulou A, Martin F and Ross A J 2001 A full description of potential curve of the B $^1\Pi_u$ state of the state of ^{7}Li$_2$ *J. Chem. Phys.* **114** 8445

[52] Bouloufa N, Cacciani P, Kokoouline V, Masnou-Seeuws F, Vetter R and Li L 2001 Predissociation induced by ungerade-gerade symmetry breaking in the $B\,{}^1\Pi_u$ state of the ${}^6Li{}^7Li$ molecule *Phys. Rev.* A **63** 042507

[53] Sinha M P, Schultz A and Zare R N 1973 Internal state distribution of alkali dimers in supersonic nozzle beams *J. Chem. Phys.* **58** 549

[54] Sinha M P, Caldwell C D and Zare R N 1974 Alignment of molecules in gaseous transport: Alkali dimers in supersonic nozzle beams *J. Chem. Phys.* **61** 491

[55] Herrmann A, Leutwyler S, Schumacher E and Wöste L 1977 Multiphoton ionization: mass selective laser-spectroscopy of Na_2 and K_2 in molecular beams *Chem. Phys. Lett.* **52** 418

[56] Mathur B P, Rothe Erhard W and Reck Gene P 1978 Two-photon ionization of Na_2 by an Ar^+ laser *J. Chem. Phys.* **68** 2518

[57] Herrmann A, Schumacher E and Wöste L 1978 Preparation and photoionization potentials of molecules of sodium, potassium, and mixed atoms *J. Chem. Phys.* **68** 2327

[58] Bergmann K, Engelhard R, Hefter U and Hering P 1979 Molecular beam diagnostics with state selection: Intensity distribution of a Na/Na_2 supersonic beam *Chem. Phys.* **44** 23

[59] Hofmann M, Leutwyler S and Schulze W 1979 Matrix isolation/aggregation of sodium atoms and molecules formed in a supersonic nozzle beam *Chem. Phys.* **40** 145

[60] Herrmann A, Leutwyler S, Wöste L and Schumacher E 1979 Molecular spectroscopy by photodeflection of Na_2 in a supersonic nozzle beam *Chem. Phys. Lett.* **62** 444

[61] Rothe E W, Krause U and Duren R 1980 Observation of polarization of atomic fluorescence excited by laser induced dissociation *Chem. Phys. Lett.* **72** 100

[62] Bergmann K and Gottwald E 1981 Effect of optical pumping in two step photoassociation of Na_2 in molecular beams *Chem. Phys. Lett.* **78** 515

[63] Kompitsas M, Kolwas K and Weber H G 1981 Relaxation in a Na/Na_2 nozzle expansion *Chem. Phys.* **55** 221

[64] Leutwyler S, Hofmann M, Harri H-P and Schumacher E 1981 The adiabatic ionization potentials of the alkali dimers Na_2, NaK and K_2 *Chem. Phys. Lett.* **77** 257

[65] Engelke F, Hage H and Caldwell C D 1982 $Na_2\ b\,{}^3\Pi_u - X\,\Sigma_g^+$ intercombination bands: Direct observation through high resolution laser fluorescence spectroscopy in a supersonic nozzle beam *Chem. Phys.* **64** 221

[66] Delacrétaz G, Ganière J D, Monot R and Wöste L 1982 Photoionization and fragmentation of alkali metal clusters in supersonic molecular beams *Appl. Phys.* B **29** 55

[67] Atkinson J B, Becker J and Demtröder W 1982 Hyperfine structure of the 625 nm band in the $a\,{}^3\Pi_u \leftarrow X\,{}^1\Sigma_g^+$ transitions of Na_2 *Chem. Phys. Lett.* **87** 128

[68] Atkinson J B, Becker J and Demtröder W 1982 Experimental observation of the $a\,{}^3\Pi_u$ state of Na_2 *Chem. Phys. Lett.* **87** 92

[69] Gole J L, Green G J, Pace S A and Preuss D R 1982 The characterization of supersonic sodium vapor expansions including laser induced atomic fluorescence from trimeric sodium *J. Chem. Phys.* **76** 2247

[70] Jones P L, Gaubatz U, Hefter U, Bergmann K and Wellegehausen B 1983 Optically pumped sodium dimer supersonic beam laser *Appl. Phys. Lett.* **42** 222

[71] Delacrétaz G and Wöste L 1985 Two-photon ionization spectroscopy of the $(2){}^1\Sigma_u^+$ double-minimum state of Na_2 *Chem. Phys. Lett.* **120** 342

[72] Gerber G and Möller R 1985 Optical-optical double-resonance spectroscopy of high vibrational levels of the $Na_2\ A\,{}^1\Sigma_u^+$ state in a molecular beam *Chem. Phys. Lett.* **113** 546

[73] Goy P, Bordas M Chr, Broyer M, Labastie P and Tribollet B 1985 Microwave transitions between molecular Rydberg states *Chem. Phys. Lett.* **120** 1

[74] Haugstätter R, Goerke A and Hertel I V 1988 Case studies in multiphoton ionisation and dissociation of Na_2 Z. Phys. D **9** 153

[75] Bordas C, Labastie P, Chevaleyre J and Broyer M 1989 MQDT analysis of rovibrational interactions and autoionization in Na_2 Rydberg states Chem. Phys. **129** 21

[76] Bordas C, Broyer M and Vialle J L 1990 Spectroscopy of the $1\ ^2\Pi_u$ state of Na_2^+ J. Chem. Phys. **92** 4030

[77] Kumar S V K, Ziegler G, Korsch H J and Bergmann K 1991 Inelastic transitions in vibrationally excited Na_2 induced by intermediate-energy-electron impact Phys. Rev. A **44** 268

[78] Richter H, Knöckel H and Tiemann E 1991 The potential of the Na_2 B $^1\Pi_u$ state Chem. Phys. **157** 217

[79] Knöckel H, Johr T, Richter H and Tiemann E 1991 The influence of the spin-orbit and the hyperfine interaction on the asymptotic behaviour of the A $^1\Sigma_u^+$ state of Na_2 Chem. Phys. **152** 399

[80] Zalicki P, Billy N, Gouedard G and Vigue J 1993 Terminal rovibrational distribution of Na_2 in a sodium supersonic beam J. Chem. Phys. **99** 6436

[81] Färbert A, Koch J, Platz T and Demtröder W 1994 Vibrationally resolved resonant two-photon ionization spectroscopy of the $1\ ^3\Sigma_g^+$ (b)$\rightarrow$$1\ ^3\Sigma_u^+$ (X) transition of Na_2 Chem. Phys. Lett. **223** 546

[82] Elbs M, Knöckel H, Laue T, Samuelis C and Tiemann E 1999 Observation of the last bound levels near the Na_2 ground-state asymptote Phys. Rev. A **59** 3665

[83] Samuelis Chr, Falke St, Laue T, Pellegrini P, Dulieu O, Knöckel H and Tiemann E 2003 Optical manipulation of long-range interactions at the 3s+3p asymptote of Na_2 Eur. Phys. J. D **26** 307

[84] Laue T, Pellegrini P, Dulieu O, Samuelis Chr, Knöckel H and Tiemann E 2003 Observation of the long-range potential well of the $(6)^1\Sigma_g^+$ (3s+5s) state of Na_2 Eur. Phys. J. D **26** 173

[85] Ghazy R, Hamada I M, Demtröder W, El-Kashef H and Hassan G E 2004 Sub-Doppler laser spectroscopy of Na_2 in a cold molecular beam Egypt. J. Sol. **27** 77

[86] Garcia-Fernandez R, Ekers A, Klavins J, Yatsenko L P, Bezuglov N N, Shore B W and Bergmann K 2005 Autler-Townes effect in a sodium molecular-ladder scheme Phys. Rev. A **71** 023401

[87] Leutwyler S, Hermann A, Wöste L and Schumacher E 1980 Isotope selective two-step photoionization study of K_2 Chem. Phys. B **48** 253

[88] Meiwes K H and Engelke F 1982 Predissociation of K_2: molecular beam-laser-induced fluorescence spectroscopy of the C $^1\Pi_u$ – X $^1\Sigma_g^+$ band system Chem. Phys. Lett. **85** 409

[89] Heinze J, Kowalczyk P and Engelke F 1988 Quasibound levels and shape resonances of $^{39}K_2$ (B $^1\Pi_u$) crossed laser-molecular beam studies and analytical interpretation J. Chem. Phys. **89** 3428

[90] Kowalczyk P, Schühle U and Engelke F 1989 Rydberg states of the K_2 molecule studied by laser spectroscopy in a supersonic beam Z. Phys. D **13** 231

[91] Joo D, Yoon Y, Lee Y, Baek S and Kim B 2000 New electric quadrupole transitions of K_2 observed in a pulsed molecular beam: The $1\ ^1\Delta_g$ – $X^1\Sigma_g^+$ bands near 500 nm J. Chem. Phys. (Communication) **113** 2945

[92] Lisdat C, Knöckel H and Tiemann E 2000 First Observation of Hyperfine Structure in K_2 J. Mol. Spectrosc. **199** 81

[93] St. Falke I, Sherstov E, Tiemann and Lisdat Ch 2006 The A $^1\Sigma_u^+$ state of K_2 up to the dissociation limit J. Chem. Phys. **125** 224303

[94] Sherstov I, Liu S, Lisdat Ch, Schnatz H, Jung S, Knöckel H and Tiemann E 2007 Frequency measurements in the $b\,^3\Pi^+_{0u} - X^1\Sigma_g^+$ system of K$_2$ *Eur. Phys. J.* D **41** 485

[95] Falke S, Knöckel H, Friebe J, Riedmann M and Tiemann E 2008 Potassium ground-state scattering parameters and Born-Oppenheimer potentials from molecular spectroscopy *Phys. Rev.* A **78** 012503

[96] Liu S, Sherstov I, Lisdat C, Knöckel H and Tiemann E 2010 Ramsey-Bordé interferometer and embedded Ramsey interferometer with molecular matter waves of ^{39}K$_2$ *Eur. Phys. J.* D **58** 369

[97] Lee Y, Yoon Y, Baek S J, Joo D-L, Ryu J and Kim B 2000 Direct observation of the $2\,^3\Pi_u$ state of Rb$_2$ in a pulsed molecular beam: Rotational branch intensity anomalies in the $2\,^3\Pi_u(1_u) - X\,^1\Sigma_g^+(0_g^+)$ bands *J. Chem. Phys.* **113** 2116

[98] Breford E J and Engelke F 1980 Laser-induced fluorescence in supersonic nozzle beams: predissociation in the Rb$_2$ $C\,^1\Pi_u$ and $D\,^1\Pi_u$ states *Chem. Phys. Lett.* **75** 132

[99] Caldwell C D, Engelke F and Hage H 1980 High resolution spectroscopy in supersonic nozzle beams: The Rb$_2$ $B\,^1\Pi_u - X\,^1\Sigma_g^+$ band system *Chem. Phys.* **54** 21

[100] Yoon Y, Lee Y, Lee S and Kim B 2002 Electric quadrupole transitions of Rb$_2$ observed in a pulsed molecular beam: The $1\,^1\Delta_g - X^1\Sigma_g^+$ bands near 540 nm *J. Chem. Phys.* **116** 6660

[101] Lee Y, Lee S and Kim B 2007 Mass-resolved resonance enhanced ionization study of complicated excited electronic states of Rb$_2$ near 430 nm and their predissociation dynamics *J. Phys. Chem.* A **111** 11750

[102] Lee Y, Lee S and Kim B 2008 Spin-forbidden transitions in the vicinity of the $2\,^1\Pi_u \leftarrow X\,^1\Sigma_g^+$ band system of Rb$_2$ *J. Phys. Chem.* A **112** 6893

[103] Höning G, Czajkowski M, Stock M and Demtröder W 1979 High resolution laser spectroscopy of Cs$_2$: I. ground state constants and potential curve *J. Chem. Phys.* **71** 2138

[104] Katô H and Yoshihara K 1979 Laser induced fluorescence, energy transfer and dissociation of Cs$_2$ *J. Chem. Phys.* **71** 1585

[105] Kim B and Yoshihara K 1993 Determination of adiabatic ionization potentials of Cs$_2$ and Cs$_3$ in a very cold molecular beam using time-of-flight mass spectroscopy *Chem. Phys. Lett.* **202** 437

[106] Kim B and Yoshihara K 1993 Triplet-triplet transition of Cs$_2$ studied by multiphoton ionization spectroscopy in a very cold pulsed molecular beam *Chem. Phys. Lett.* **204** 407

[107] Kim B and Yoshihara K 1993 The 480 nm system of Cs$_2$ studied in a very cold molecular beam: Direct observation of a new E″ and the ion-pair states *J. Chem. Phys.* **98** 5990

[108] Kim B 1993 Direct observation of the (2) $^3\Pi_u$ state of Cs$_2$ by resonance enhanced two photon ionization spectroscopy in a very cold molecular beam *J. Chem. Phys.* **99** 5677

[109] Kim B, Yoshihara K and Lee S Y 1994 Complex resonances in the predissociation of Cs$_2$ *Phys. Rev. Lett.* **73** 424

[110] Lee Y 2002 Excited-state spectroscopy of Rb$_2$ and KRb by resonance enhanced two photon ionization in a supersonic molecular beam Ph. D. Thesis, KAIST, Korea

[111] Kimura Y, Lefebvre-Brion H, Kasahara S, Katô H, Baba M and Lefebvre R 2000 Interference effects in the predissociation of the Cs$_2$ $C\,^1\Pi_u$ and (2) $^3\Pi_u$ states through the dissociative (2) $^3\Sigma_u^+$ state *J. Chem. Phys.* **113** 8637

[112] Bouloufa N, Cacciani P, Vetter R and Yiannopoulou A 2000 Sub-Doppler spectroscopy of the LiH molecule: The $A - X$ system *J. Mol. Spectrosc.* **202** 37

[113] Freeman R R, Jacobson A R, Johnson D W and Ramsey N F 1975 The molecular Zeeman and hyperfine spectra of LiH and LiD by molecular beam high resolution electric resonance *J. Chem. Phys.* **63** 2597

[114] Dagdigian P J 1976 Detection of LiH and NaH molecular beams by laser fluorescence and measurement of radiative lifetimes of the $A\ ^1\Sigma^+$ state *J. Chem. Phys.* **64** 2609

[115] Dagdigian P J and Wharton L 1972 Molecular beam electric deflection and resonance spectroscopy of the heteronuclear alkali dimers: $^{39}K^7Li$, Rb^7Li, $^{39}K^{23}Na$, $Rb^{23}Na$, and $^{133}Cs^{23}Na$ *J. Chem. Phys.* **57** 1487

[116] Pesl F P, Lutz S and Bergmann K 2000 Improved molecular constants for the $X\ ^1\Sigma^+$ and $A\ ^1\Sigma^+$ states of NaH *Eur. Phys. J.* D **10** 247

[117] Breford E J and Engelke F 1978 Laser-induced molecular fluorescence in supersonic nozzle beams: Applications to the NaK $D\ ^1\Pi - X\ ^1\Sigma^+$ and $D\ ^1\Pi - a\ ^3\Sigma^+$ system *Chem. Phys. Lett.* **53** 282

[118] Gerdes A, Dulieu O, Knöckel H and Tiemann E 2011 Stark effect measurements on the NaK molecule *Eur. Phys. J* D **65** 105

[119] Lee Y, Yun C, Yoon Y, Kim T and Kim B 2001 The 530 nm system of KRb observed in a pulsed molecular beam: New electric quadrupole transitions ($1\ ^1\Delta - X^1\Sigma^+$) *J. Chem. Phys.* **115** 7413

[120] Lee Y, Yoon Y, Kim B, Li L and Lee S 2004 Observation of the $3\ ^3\Sigma^+ - X\ ^1\Sigma^+$ transition of KRb by resonance enhanced two-photon ionization in a pulsed molecular beam: Hyperfine structures of $^{39}K^{85}Rb$ and $^{39}K^{87}Rb$ isotopomers *J. Chem. Phys.* **120** 6551

[121] Lee Y, Yoon Y, Muhammad A, Kim J T, Lee S and Kim B 2010 The 480 nm system of KRb: The $1\ ^3\Delta_1$, $4\ ^1\Sigma^+$ and $5\ ^1\Sigma^+$ states *J. Phys. Chem.* A **114** 7742

[122] Lee Y, Yoon Y, Kim J T, Lee S and Kim B 2011 Unravelling complex spectra of a simple molecule: REMPI study of 420 nm system of KRb *Chem. Phys. Chem.* **12** 2018

[123] Kim B and Yoshihara K 1993 $^3\Delta - ^1\Sigma^+$ transition of RbCs observed in a very cold molecular beam *Chem. Phys. Lett.* **212** 271

[124] Kim B and Yoshihara K 1994 Resonance enhanced two photon ionization spectroscopy of RbCs in a very cold molecular beam *J. Chem. Phys.* **100** 1849

[125] Yoon Y, Lee Y, Kim T, Ahn J S, Jung Y, Kim B and Lee S 2001 High resolution resonance enhanced two photon ionization spectroscopy of RbCs in a cold molecular beam *J. Chem. Phys.* **114** 8926

[126] Lee Y, Yoon Y, Lee S, Kim J T and Kim B 2008 Parallel and coupled perpendicular transitions of RbCs 640 nm system: mass-resolved resonance enhanced two-photon ionization in a cold molecular beam *J. Phys. Chem.* A **112** 7214

[127] Lee Y, Yoon Y, Lee S and Kim. B 2009 500 nm System of RbCs: Assignments and Intensity Anomalies *J. Phys. Chem.* A **113** 12187

[128] Barry J F, Schman E S, Norrgard E B and DeMille D 2012 Laser radiation pressure slowing of a molecular beam *Phys. Rev. Lett.* **108** 103002

[129] Davis K B, Mewes M-O, Andrews M R, van Druten N J, Durfee D S, Kurn D M and Ketterle W 1995 Bose-Einstein condensation in a gas of sodium atoms *Phys. Rev. Lett.* **75** 3969

[130] Balewski J B, Krupp A T, Gaj A, Peter D, Büchler H P, Löw R, Hofferberth S and Pfau T 2013 2D magneto-optical trapping of diatomic molecules *Nature* **502** 664

[131] McAlexander W I, Abraham E R I and Ritchie N W M 1995 Precise atomic radiative lifetime via photoassociative spectroscopy of ultracold lithium *Phys. Rev.* A **51** 871

[132] Abraham E R I, Ritchie N W M, McAlexander W I and Hulet R G 1995 Photoassociative spectroscopy of long-range states of ultracold $^6Li_2$ and $^7Li_2$ *J. Chem. Phys.* **103** 7773

[133] Lett P D, Helmerson K, Phillips W D, Ratlia L P, Rolston S L and Wagshul M E 1993 Spectroscopy of Na_2 by Photoassociation of Laser-Cooled Na *Phys. Rev. Lett.* **71** 2200

[134] Ratliff L P, Wagshul M E, Lett P D, Rolston S L and Phillips W D 1994 Photoassociative spectroscopy of 1_g, 0^+_u, and 0^-_g state of Na_2 *J. Chem. Phys.* **101** 2638

[135] Jones K M, Maleki S, Ratliff L P and Lett P D 1997 Two-color photoassociation spectroscopy of ultracold sodium *J. Phys. B: At. Mol. Opt. Phys* **30** 289

[136] Shaffer J P, Chalupczak W and Bigelow N P 1999 Highly excited states of ultracold molecules: Photoassociative spectroscopy of Na_2 *Phys. Rev. Lett.* **83** 3621

[137] Amelink A, Jones K M, Lett P D, van der Straten P and Heideman H G M 2000 Spectroscopy of autoionizing doubly excited states in ultracold Na_2 molecules produced by photoassociation *Phys. Rev. A* **61** 042707

[138] Amelink A, Jones K M, Lett P D, van der Straten P and Heideman H G M 2000 Single-color photoassociative ionization of ultracold sodium: The region from 0 to 5 GHz *Phys. Rev. A* **62** 013408

[139] Wang H, Wang X T, Gould P L and Stwalley W C 1997 Optical-optical double resonance photoassociative spectroscopy of ultracold ^{39}K atoms near highly excited asymptotes *Phys. Rev. Lett.* **78** 4173

[140] Wang X, Wang H, Gould P L and Stwalley W C 1998 Observation of the pure long-range 1_u state of an alkali-metal dimer by photoassociative spectroscopy *Phys. Rev. Lett.* **57** 4600

[141] Wang H, Li J, Wang X T, Williams C J, Gould P L and Stwalley W C 1997 Precise determination of the dipole matrix element and radiative lifetime of the ^{39}K 4p state by photoassociative spectroscopy *Phys. Rev. A* **55** 1569

[142] Wang H, Gould P L and Stwalley W C 1997 Long-range interaction of the $^{39}K(4s)+^{39}K(4p)$ asymptote by photoassociative spectroscopy. I. The 0_g^- pure long-range state and the long-range potential constant *J. Chem. Phys.* **106** 7899

[143] Wang H, Gould P L and Stwalley W C 1998 Fine-structure predissociation of ultracold photoassociated $^{39}K_2$ Molecules observed by fragmentation spectroscopy *Phys. Rev. Lett.* **80** 476

[144] Nikolov A N, Eyler E E, Wang X T, Li J, Wang H, Stwalley W C and Gould P L 1999 Observation of ultracold ground-state potassium molecules *Phys. Rev. Lett.* **82** 703

[145] Pichler M, Chen H M, Wang H, Stwalley W C, Ross A J, Martin F, Aubert-Frécon M and Russier-Antoine I 2003 Photoassociation of ultracold K atoms: Observation of high lying levels of the $1_g \sim 1$ $^1\Pi_g$ molecular state of K_2 *J. Chem. Phys.* **118** 7837

[146] Miller J D, Cline R A and Heinzen D J 1993 Photoassociation spectrum of ultracold atoms *Phys. Rev. Lett.* **71** 2204

[147] Cline R A, Miller J D and Heinzen D J 1994 Study of Rb_2 Long-range states by high-resolution photoassociation spectroscopy *Phys. Rev. Lett.* **73** 632

[148] Leonhardt D and Weiner J 1995 Direct two-color photoassociative ionization in a rubidium magneto-optic trap *Phys. Rev. A* **52** 4332

[149] Tsai C C, Freeland R S, Vogels J M, Boesten H M J M, Verhaar B J and Heinzen D J 1997 Two-color photoassociation spectroscopy of ground state Rb_2 *Phys. Rev. Lett.* **79** 1245

[150] Gabbanini C, Fioretti A, Lucchesini A, Gozzini S and Mazzoni M 2000 Cold rubidium molecules formed in a magneto-optical trap *Phys. Rev. Lett.* **84** 2814

[151] Vogels J M, Freeland R S, Tsai C C, Verhaar B J and Heinzen D J 2000 Coupled singlet-triplet analysis of two-color cold-atom photoassociation spectra *Phys. Rev.* A **61** 043407

[152] Fioretti A, Amiot C, Dion C M, Dulieu O, Mazzoni M, Smirne G and Gabbanini C 2001 Cold rubidium molecule formation through photoassociation: A spectroscopic study of the 0_g^- long range state of $^{87}Rb_2$ *Eur. J. Phys.* D **15** 189

[153] Kemmann M, Mistrik I, Nussmann S, Helm H, Williams C J and Julienne P S 2004 Near-threshold photoassociation of $^{87}Rb_2$ *Phys. Rev.* A **69** 022715

[154] Huang Y, Qi J, Pechkis H K, Wang D, Eyler E E, Gould P L and Stwalley W C 2006 Detection by two-photon ionization and magnetic trapping of cold Rb_2 triplet state molecules *J. Phys. B: At. Mol. Opt. Phys.* **39** S857

[155] Lozeille J, Fioretti A, Gabbanini C, Huang Y, Pechkis H K, Wang D, Gould P L, Eyler E E, Stwalley W C, Aymar M and Dulieu O 2006 Detection by two-photon ionization and magnetic trapping of cold Rb_2 triplet state moleculues *Eur. J. Phys.* D **39** 261

[156] Ye Huang 2006 Production, Detection and Trapping of Ultracold Molecular Rubidium Ph. D. Thesis, Univ. of Connecticut, USA

[157] Dulieu O Private communication

[158] Bellos M A, Rahmlow D, Carollo R, Banerjee J, Dulieu O, Gerdes A, Eyler E E, Gould P L and Stwalley W C 2011 Formation of ultracold Rb_2 molecules in the $v'' = 0$ level of the $a^3\Sigma_u^+$ state via blue-detuned photoassociation to the $1\,^3\Pi_g$ state *Phys. Chem. Chem. Phys.* **13** 18880

[159] Bellos M A, Carollo R, Banerjee J, Ascoli M, Allouche A-R, Eyler E E, Gould P L and Stwalley W C 2013 Upper bound to the ionization energy of $^{85}Rb_2$ *Phys. Rev.* A **87** 012508

[160] Carollo R, Bellos M A, Rahmlow D, Banerjee J, Eyler E E, Gould P L and Stwalley W C 2013 Observation and analysis of resonant coupling between nearly degenerate levels of the $2\,^1\Sigma_g^+$ and $1\,^1\Pi_g$ states of ultracold $^{85}Rb_2$ *Phys. Rev.* A **87** 022505

[161] Fioretti A, Comparat D, Drag C, Amiot C, Dulieu O, Masnou-Seeuws F and Pillet P 1999 Photoassociative spectroscopy of the Cs_2 0_g^- long-range state *Eur. J. Phys.* D **5** 389

[162] Comparat D, Drag C, Laburthe Tolra B, Fioretti A, Pillet P, Crubellier A, Dulieu O and Masnou-Seeuws F 2000 Formation of cold Cs_2 ground state molecules through photo-association in the 1_u pure long-range state *Eur. J. Phys.* D **11** 59

[163] Dion C M, Drag C, Dulieu O, Laburthe Tolra B, Masnou-Seeuws F and Pillet P 2001 Resonant coupling in the formation of ultracold ground state molecules via photo-association *Phys. Rev. Lett.* **86** 2253

[164] Pichler M, Chen H and Stwalley W C 2004 Photoassociation spectroscopy of ultracold Cs below the $6P_{3/2}$ limit *J. Chem. Phys.* **121** 6779

[165] Pichler M, Chen H and Stwalley W C 2004 Photoassociation spectroscopy of ultracold Cs below the $6P_{1/2}$ limit *J. Chem. Phys.* **121** 1796

[166] Pichler M, Qi J, Stwalley W C, Beuc R and Pichler G 2006 Observation of blue satellite bands and photoassociation at ultracold temperatures *Phys. Rev.* A **73** 021403

[167] Pichler M, Stwalley W C and Dulieu O 2006 Perturbation effects in photoassociation spectra of ultracold Cs_2 *J. Phys. B: At. Mol. Opt. Phys.* **39** S981

[168] Danzl J G, Haller E, Gustavsson M, Mark M J, Hart R, Bouloufa N, Dulieu O, Ritsch H and Nägerl H-C 2008 Quantum gas of deeply bound ground state molecules *Science* **321** 1062

[169] Viteau M, Chotia A, Allegrini M, Bouloufa N, Dulieu O, Comparat D and Pillet P 2009 Efficient formation of deeply bound ultracold molecules probed by broadband detection *Phys. Rev.* A **79** 021402(R)

[170] Ma J, Wang L, Zhao Y, Xiao L and Jia S 2009 Direct Measurement of the Cs_2 0_u^+ ($P_{3/2}$) High-lying vibrational spectroscopy using photon counting *J. Phys. Soc. Jpn* **78** 064302

[171] Danzl J G, Mark M J, Haller E, Gustavsson M, Hart R, Aldegunde J, Hutson J M and Nägerl H-C 2010 An ultracold high-density sample of rovibronic ground-state molecules in an optical lattice *Nature Phys.* **6** 265

[172] Ridinger A, Chaudhuri S, Salez T, Fernandes D R, Bouloufa N, Dulieu O, Salomon C and Chevy F 2011 Saturation in heteronuclear photoassociation of ^{6}Li^{7}Li *Europhys. Lett.* **96** 33001

[173] Deiglmayr J, Grochola A, Repp M, Mörtlbauer K, Glück C, Lange J, Dulieu O, Wester R and Weidemüller M 2008 Photoassociative creation of ultracold heteronuclear ^{6}Li^{40}K* molecules *Phys. Rev. Lett.* **101** 133004

[174] Wang H and Stwalley W C 1998 Ultracold photoassociative spectroscopy of heteronuclear alkali-metal diatomic molecules *J. Chem. Phys.* **108** 5767

[175] Wang D, Qi J, Stone M F, Nikolayeva O, Hattaway B, Gensemer S D, Wang H, Zemke W T, Gould P L, Eyler E E and Stwalley W C 2004 The photoassociative spectroscopy, photoassociative molecule formation, and trapping of ultracold ^{39}K^{85}Rb *Eur. Phys. J.* D **31** 165

[176] Wang D, Qi J, Stone M F, Nikolayeva O, Wang H, Hattaway B, Gensemer S D, Gould P L, Eyler E E and Stwalley W C 2004 Photoassociative production and trapping of ultracold KRb molecules *Phys. Rev. Lett.* **93** 243005

[177] Wang D, Eyler E E, Gould P L and Stwalley W C 2006 Spectra of ultracold KRb molecules in near-dissociation vibrational levels *J. Phys. B: At. Mol. Opt. Phys.* **39** S849

[178] Wang D, Kim J T, Ashbaugh C, Eyler E E, Gould P L and Stwalley W C 2007 Rotationally resolved depletion spectroscopy of ultracold KRb molecules *Phys. Rev. A* **75** 032511

[179] Kim J T, Wang D, Eyler E E, Gould P L and Stwalley W C 2009 Spectroscopy of ^{39}K^{85}Rb triplet excited states using ultracold a $^3\Sigma^+$ state molecules formed by photoassociation *New J. Phys.* **11** 055020

[180] Stwalley W C, Banerjee J, Bellos M, Carollo R, Recore M and Mastroianni M 2010 Resonant coupling in the heteronuclear alkali dimers for direct photoassociative formation of $X(0,0)$ ultracold molecules *J. Phys. Chem. A* **114** 81

[181] Kim J T, Lee Y, Kim B, Wang D, Stwalley W C, Gould P L and Eyler E E 2011 Spectroscopic analysis of the coupled 1 $^1\Pi$, 2 $^3\Sigma^+$) ($\Omega = 0^-$, 1), and b $^3\Pi$ ($\Omega = 0^\pm$, 1, 2) states of the KRb molecule using both ultracold molecule and molecular beam experiment *Phys. Chem. Chem. Phys.* **13** 18755

[182] Kim J T, Lee Y, Kim B, Wang D, Gould P, Eyler E and Stwalley W 2012 Spectroscopic investigation of the A and 3 $^1\Sigma^+$ states of ^{39}K^{85}Rb *J. Chem. Phys.* **137** 244301

[183] Banerjee J, Rahmlow D, Carollo R, Bellos M, Eyler E E, Gould P L and Stwalley W C 2013 Spectroscopy and applications of the 3 $^3\Sigma^+$ electronic state of ^{39}K^{85}Rb *J. Chem. Phys.* **139** 174316

[184] Banerjee J, Rahmlow D, Carollo R, Bellos M, Eyler E E, Gould P L and Stwalley W C 2013 Spectroscopy of the double minimum 3 $^3\Pi_\Omega$ electronic state of ^{39}K^{85}Rb *J. Chem. Phys.* **138** 164302

[185] Ospelkaus S, Pe'er A, Ni K-K, Zirbel J J, Neyenhuis B, Kotochigova S, Julienne P S, Ye J and Jin D S 2008 Efficient state transfer in an ultracold dense gas of heteronuclear molecules *Nature Phys.* **4** 622

[186] Viteau M, Chotia A, Allegrini M, Bouloufa N, Dulieu O, Comparat D and Pillet P 2008 Optical pumping and vibrational cooling of molecules *Science* **321** 232

[187] Aikawa K, Akamatsu D, Hayashi M, Kobayashi J, Ueda M and Inouye S 2011 Predicting and verifying transition strengths from weakly bound molecules *Phys. Rev. A* **83** 042706

[188] Zabawa P, Wakim A, Haruza M and Bigelow N P 2011 Formation of ultracold $X\,^1\Sigma^+(v''=0)$ NaCs molecules via coupled photoassociation channels *Phys. Rev. A* **84** 061401

[189] Haimberger C, Kleinert J, Bhattacharya M and Bigelow N P 2004 Formation and detection of ultracold ground-state polar molecules *Phys. Rev. A* **70** 021402

[190] Haimberger C, Kleinert J, Zabawa P, Wakim A and Bigelow N P 2009 Formation of ultracold, highly polar $X\,^1\Sigma^+$ NaCs molecules *New J. Phys.* **11** 055042

[191] Zabawa P, Wakim A, Neukirch A, Haimberger C, Bigelow N P, Stolyarov A V, Pazyuk E A, Tamanis M and Ferber R 2010 Near-dissociation photoassociative production of deeply bound NaCs molecules *Phys. Rev. A* **82** 040501

[192] Wakim A, Zabawa P and Bigelow N P 2011 Photoassociation studies of ultracold NaCs from the Cs $6\,^2P_{3/2}$ asymptote *Phys. Chem. Chem. Phys.* **13** 18887

[193] Bergeman T, Kerman A J, Sage J, Sainis S and DeMille D 2004 Prospects for production of ultracold $X\,^1\Sigma^+$ RbCs molecules *Eur. J. Phys. D* **31** 179

[194] Kerman Andrew J, Sage Jeremy M, Sainis S, Bergeman T and DeMille D 2004 Production and state-selective detection of ultracold RbCs molecules *Phys. Rev. Lett.* **92** 153001

[195] Kerman A J, Sage J M, Sainis S, Bergeman T and DeMille D 2004 Production of ultracold, polar RbCs* molecules via photoassociation *Phys. Rev. Lett.* **92** 033004

[196] Sage J M, Sainis S, Bergeman T and DeMille D 2005 Optical production of ultracold polar molecules *Phys. Rev. Lett.* **94** 203001

[197] Gabbanini C and Dulieu O 2011 Formation of ultracold metastable RbCs molecules by short-range photoassociation *Phys. Chem. Chem. Phys.* **13** 18905

[198] Debatin M, Takekoshi T, Rameshan R, Reichsöllner L, Ferlaino F, Grimm R, Vexiau R, Bouloufa N, Dulieu O and Nägerl H-C 2011 Molecular spectroscopy for ground-state transfer of ultracold RbCs molecule *Phys. Chem. Phys. Chem* **13** 18926

[199] Ji Z, Zhang H, Wu J, Yuan J, Yang Y, Zhao Y, Ma J, Wang L, Xiao L and Jia S 2012 Photoassociative formation of ultracold RbCs molecules in the (2) $^3\Pi$ state *Phys. Rev. A* **85** 013401

[200] Bouloufa-Maafa N, Aymar M, Dulieu O and Gabbanini C 2012 Formation of ultracold RbCs molecules by photoassociation *Laser Phys.* **22** 1502

[201] Bruzewicz C D, Gustavsson M, Shimasaki T and DeMille D 2013 Continuous formation of vibronic ground state RbCs molecules via photoassociation *New J. Phys.* **16** 023018

[202] Fioretti A and Gabbanini C 2013 Experimental study of the formation of ultracold RbCs molecule by short-range photoassociation *Phys. Rev. A* **54** 054701

[203] Feshbach H 1962 A unified theory of nuclear reactions. II *Annals of Phys.* **19** 287

[204] Heo M-S, Wang T T, Christensen C A, Rvachov T M, Cotta D A, Choi J-H, Lee Y-R and Ketterle W 2012 Formation of ultracold fermionic NaLi Feshbach molecules *Phys. Rev. A* **86** 021602(R)

[205] Köhler T, Góral K and Julienne P 2006 Production of cold molecules via magnetically tunable Feshbach resonances *Rev. Mod. Phys.* **78** 1311

[206] Aikawa K, Akamatsu D, Hayashi M, Oasa K, Kobayashi J, Naidon P, Kishimoto T, Ueda M and Inouye S 2010 Coherent transfer of photoassociated molecules into the rovibrational ground state *Phys. Rev. Lett.* **105** 203001

[207] Beuc R, Movre M, Ban T, Pichler G, Aymar M, Dulieu O and Ernst W E 2006 Predictions for the observation of KRb spectra under cold conditions *J. Phys. B: At. Mol. Opt. Phys.* **39** S1191

[208] Banerjee J, Rahmlow D, Carollo R, Bellos M, Eyler E E, Gould P L and Stwalley W C 2012 Direct photoassociative formation of ultracold KRb molecules in the lowest vibrational levels of the electronic ground state *Phys. Rev. A* **86** 053428

[209] Birzniece I, Nikolayeva O, Tamanis M and Ferber R 2012 $B(1)$ $^1\Pi$ state of KCs: High-resolution spectroscopy and description of low-lying energy levels *J. Chem. Phys.* **136** 064304

[210] Docenko O, Tamanis M, Ferber R, Pazyuk E A, Zaitsevskii A, Stolyarov A V, Pashov A, Knöckel H and Tiemann E 2007 Deperturbation treatment of the A $^1\Sigma^+ - b^3\Pi$ complex of NaRb and prospects for ultracold molecule formation in X $^1\Sigma^+$ ($v = 0$; $J = 0$) *Phys. Rev. A* **75** 042503

[211] Kim J T, Lee Y, Kim B, Wang D, Stwalley W C, Gould P L and Eyler E E 2011 Spectroscopic prescription for optimal stimulated Raman transfer of ultracold heteronuclear molecules to the lowest rovibronic level *Phys. Rev. A* **84** 062511

[212] Kim J T 2013 Population transfer routes to the lowest vibrational level of ultracold $^{39}K^{85}Rb$ *J. Kor. Phys. Soc.* **63** 933

[213] Rousseau S, Allouche A R and Aubert-Frécon M 2000 Theoretical study of the electronic structure of the KRb molecule *J. Mol. Spectrosc.* **203** 235

IOP Concise Physics

Analysis of the Alkali Metal Diatomic Spectra
Using molecular beams and ultracold molecules
Jin-Tae Kim, Bongsoo Kim and William C Stwalley

Chapter 2

Experimental methods

2.1 Review of previous supersonic MB experiments

There are many previous publications [1–6] which have used a supersonic MB, but we focus on the alkali molecule publications [1, 6]. The first continuous supersonic MB machine [7] for condensable gases (which was composed of three main parts with nozzle, skimmer, and collimator) was constructed by modifying a conventional effusive beam machine used for atomic beam magnetic resonance experiments [8]. The supersonic MB formed through adiabatic gas expansion at a nozzle passed through a skimmer followed by a collimation slit to produce the MB.

MB kinematics, velocity distributions, recombination rate constants, internal state distributions [1, 3, 5, 9], etc of alkali molecules using the supersonic beams operated at high temperature have been extensively investigated [2, 3, 10–93]. Internal energy cooling of molecules concentrates the population in low vibrational and rotational levels of the ground state. Finally, a great reduction in spectral complexity due to internal cooling of the NO_2 molecule made it possible to assign individual rovibrational levels in absorption spectra for the first time [1].

Although continuous supersonic MBs produced significant internal cooling of the molecule, still more internal cooling was desirable to reduce the population in higher vibrational and rotational levels in the ground electronic state of the molecule with very small rotational and vibrational constants. The pulsed MB method, which gives more efficient translational cooling, a more intense beam, a higher concentration, and less stresses in the pumping system, has more internal cooling compared to those of continuous supersonic MBs.

The development of pulsed lasers with peak powers high enough to ionize molecules also led to such pulsed MBs. There have been many pulsed supersonic MB developments [94–101]. The first pulsed supersonic MB of a few hundred microseconds pulse width [96] was used to characterize viscous regions of the MB system. Since that development, supersonic pulses with several microseconds to

doi:10.1088/978-1-6270-5678-6ch2

ms pulse widths and several Hz pulse repetitions [97, 98] have been developed. Various pulse generation methods using a solenoid [98], discharge of a capacitor [94], the magnetic repulsion of a spring bar [100], or a fuel injector valve [95, 101], have been developed. Pulsed laser development also made these supersonic pulses useful in obtaining simplified spectra.

In our experiment a supersonic pulsed nozzle was made by modifying a fuel injector [57, 63, 71, 89, 92] of an automobile (Nippon Denso). However, alkali metals have high evaporation temperatures so that the nozzle at high temperature is cooled using water cooling. A pulsed MB [71, 89] gives colder internal vibrational and rotational temperatures of $T_v \sim 5\,\mathrm{K}$ and $T_r \sim 1\,\mathrm{K}$, respectively, in the case of Cs_2. High resolution spectroscopic analysis with simplified spectra due to such cold internal temperatures of the alkali molecules were not investigated well until the pulsed supersonic MB techniques were successfully applied to alkali molecules such as Cs_2 [71], RbCs [89], and Rb_2 [63]. KRb studies are described in detail in section 3.1.

Since 2000, heteronuclear alkali dimers such as RbCs, as well as homonuclear alkali dimers such as K_2 and Rb_2, have been investigated by RE2PI in a very cold supersonic MB. This method has proven its strength in the detection of very weak transitions, including the electric quadrupole $1\,^1\Delta_g - X\,^1\Sigma_g^+$ bands of K_2 near 500 nm [57], and the electric quadrupole transitions of Rb_2 observed in a supersonic pulsed MB [66]. High resolution RE2PI studies of Rb_2 and RbCs revealed quantum interference effects in the Rb_2 $2\,^3\Pi_u \leftarrow X\,^1\Sigma_g^+$ transitions with rotational branch intensity anomalies in the $2\,^3\Pi_u(1_u) \leftarrow X\,^1\Sigma_g^+(0_g^+)$ bands, and the potential energy curve (PEC) of the RbCs $5\,^1\Sigma^+$ state [21]. Very complicated $2\,^1\Pi_u \leftarrow X\,^1\Sigma_g^+$ spectra of Rb_2 near 475 nm with spin-forbidden transitions [68], complicated excited electronic states of Rb_2 near 430 nm and their predissociation dynamics [67], and analysis of the spectra with parallel and coupled perpendicular transitions of RbCs near 640 nm [92] have been obtained and the contributing excited electronic states identified. In the Rb_2 430 nm band system, fine structure predissociation channels have also been identified by simultaneously obtained photofragment yield spectra of ^{85}Rb and ^{87}Rb. In the Rb_2 475 nm band system, Franck–Condon (FC) dark $2\,^3\Pi_u(1_u)$ and $3\,^3\Sigma_u^+(1_u)$ states have been identified. In the RbCs 640 nm band system, three coupled perpendicular transitions $(2\,^1\Pi_1 - 2\,^3\Pi_1 - 3\,^3\Sigma_1^+ \leftarrow X\,^1\Sigma_g^+)$ and one regular parallel transition $(2\,^3\Pi_0 \leftarrow X\,^1\Sigma_g^+)$ have been identified. The RKR PEC of the $2\,^3\Pi_0$ state has been constructed.

2.2 Our supersonic MB experimental methods

We used a pulsed supersonic MB to produce alkali metal vapors instead of a CW supersonic MB. This technique significantly cools the rotational and vibrational temperatures. Thus, the excitation spectra are greatly simplified and the intensity increased. We recorded excitation spectra of KRb in lower vibrational levels ($v'' = 0$ and 1) of the ground state by RE2PI.

MBs of very dense KRb molecules were produced by expanding the mixed alkali metal vapors (K and Rb) with Kr inert gas using a homemade high-temperature pulsed nozzle. The mixed metal alkali KRb diatomic molecules such as $^{39}K^{85}Rb$,

^{39}K^{87}Rb, ^{41}K^{85}Rb, and ^{41}K^{87}Rb are produced. Those isotopologues can be separated by mass in a time-of-flight (TOF) mass spectrometer (although ^{39}K^{87}Rb and ^{41}K^{85}Rb cannot easily be separated due to their very similar masses).

The pulsed supersonic nozzle is made of a modified fuel injector of an automobile (Nippon Denso). The nozzle temperature can be raised up to 500 °C. However, alkali metals have high evaporation temperatures so that the nozzle at high temperature is cooled down with water cooling. The temperature of the nozzle was maintained at about 340 °C to produce sufficient vaporization of K and Rb metals in this experiment. The nozzle was opened and closed with a solenoidal valve. In order to obtain spectra at high rotational temperatures, the jet is expanded at lower stagnation pressures of Ar, Ne, or He gases. The pressures of seeded inert gases can be varied from 760 Torr to several tens of Torr. Internal kinetic energies of metal vapors expanding adiabatically through the nozzle of an 800 μm diameter are lowered due to the collisions between the metal vapors and the inert gases as alkali diatomic molecules are generated. The translational, vibrational, and rotational temperatures of the alkali molecules can be varied by changing the pressure of the seeded inert gases.

The pulsed supersonic jet is collimated by a skimmer with a diameter of 1.2 mm (Beam Dynamics) located a few cm from the nozzle. The skimmed MB is intersected at right angles by the excitation laser in the laser–molecule interaction region of a TOF mass spectrometer. The pulsed nozzle is operated at 5 Hz.

Mass spectrometric detection using a TOF mass spectrometer is combined with RE2PI. Excitation spectra of isotopologues and fragmented products are obtained simultaneously. The absolute vibrational quantum numbers are assigned based on the isotopic shifts of vibronic bands. Also, direct and indirect dissociation dynamics can be identified. The rotational temperature can be controlled from 0.7 to 10 K. Initially, alkali dimers are mainly in the X $^1\Sigma^+$ $v'' = 0$ level. A considerable population in the X $^1\Sigma^+$ $v'' = 1$ level can be obtained. The observation of hot bands confirms the initial level of transitions.

The supersonic MB apparatus [61, 86, 92] is composed of source and detection chambers. Typical pressures of those chambers are 1.5×10^{-4} Torr for the source chamber and 6×10^{-8} Torr for the TOF detection chamber when the seeding carrier gas is Kr and the stagnation pressure is 600 Torr.

KRb$^+$ ions formed through multiphoton ionization using a Nd:YAG pumped dye laser are accelerated by a double electrostatic field TOF lens to about 3500 V and travel through a 70 cm long field free region toward a dual microchannel plate detector. The resulting ions were detected by a linear TOF mass spectrometer. The ion signal is then amplified by a fast preamplifier and sent to boxcar integrators or a very fast digital oscilloscope board installed in a PC. The ion signals with the mass-to-charge ratio $m/z = 124$ for ^{39}K^{85}Rb$^+$ were well resolved from other isotopologues of ^{39}K^{87}Rb$^+$ and ^{41}K^{85}Rb$^+$ with small natural abundances.

Excitation spectra with a low resolution of $\Delta\nu_{FWHM} \sim 0.12$ cm^{-1} are obtained using a pulsed dye laser pumped by the second or third harmonics of a Nd:YAG laser. The spectra with vibronic structures were resolved with our laser linewidth of 0.12 cm^{-1}. The narrow linewidth of $\sim$0.012 cm^{-1} of the pulsed dye laser can be obtained by installing an air spaced étalon inside the pulsed dye laser cavity. The étalon-installed cavity was pressurized to ~2.5 atm with high-purity nitrogen gas and the pressure is lowered slowly down to ~1.5 atm. Thus, spectra with rotational resolution could be obtained. Wavelength calibration is performed by the combination of a Burleigh WA-4500 wavemeter and optogalvanic spectra of neon and argon atoms. Wavelength calibrations for this high resolution scan were also made by using I_2 laser induced fluorescence spectra obtained simultaneously with the KRb molecular excitation spectra. The calibrated wavelength was accurate to within ± 0.005 cm^{-1}. For the calibration of wavelengths shorter than 500 nm, a Raman-shifted laser beam (the first Stokes line) through a pressurized H_2 pipe cell was used. The Raman-shifted frequency of 4155.226 ± 0.002 cm^{-1} was obtained when the pressure of the H_2 cell was kept constant at 11.2 atm.

2.3 Our UM experimental methods

We applied the dark-spot magneto-optical trap (MOT) method developed by Ketterle *et al* [102] to increase our trapped atom densities of the two species ^{39}K and ^{85}Rb. Attenuation of the radiation trapping force due to reabsorption of scattered photons and collisions between trapped atoms increases atom density. Thus the atoms trapped by a weak trap beam and a repumping beam with central dark spot are supplied into the central trap beam region overlapped with the central dark spot region of the repumping beam. Thus trapped atom number density is enhanced compared to a normal bright MOT without the dark spot of the repumping beam, where collisions and the radiation trapping force are greater than in the case of the dark-spot MOT. An additional depumping beam for optical pumping into the dark state was used in the case of the ^{85}Rb atom and was not used in the case of ^{39}K because of very small hyperfine splittings of the upper excited state ($4\,^2P_{3/2}$).

The repumping beam frequency of ^{85}Rb was fixed at the transition $5\,^2S\,(F''=2) \rightarrow 5\,^2P_{3/2}\,(F'=3)$ and trapping beam frequency was fixed at the transition $5\,^2S\,(F''=3) \rightarrow 5\,^2P_{3/2}\,(F'=4)$ with red detuning. The depumping beam frequency was fixed at the transition $F''=3 \rightarrow F'=3$ and reduces atom population in the bright state ($F'=3$). The trapping beam frequency of ^{39}K was fixed at the transition $F''=2 \rightarrow F'=3$ with 40 MHz red detuning. The repumper beam frequency was shifted by 462 MHz downward from the trap frequency by using an acousto-optic modulator (AOM). ^{39}K and ^{85}Rb dark-spot MOTs with densities of 3×10^{10} cm^{-3} and 1×10^{11} cm^{-3} and temperatures of 300 μK and 100 μK, respectively, have been generated and overlapped at the center of the chamber by adjusting the two MOT positions carefully, where the two different ^{39}K and ^{85}Rb atoms were trapped at slightly different positions.

Potentials in the long range regions below the K ($4S$) + Rb ($5P_{1/2}$) asymptote have large C_6 values due to van der Waals interactions [103]. Thus many levels [104, 105] of the 2(0^+), 2(0^-), 2(1), 3(0^+), 3(0^-), 3(1), 1(2), and 4(1) states below the K ($4S$) + Rb ($5P_{1/2}$) asymptote have been observed by applying a CW PA laser to colliding ultracold ^{39}K and ^{85}Rb atoms with sub-mK temperature. The molecular ion signals were obtained by detecting available molecular ions generated through the intermediate electronic states from the high vibrational levels of the $X\,^1\Sigma^+$ and $a\,^3\Sigma^+$ states formed by decay from the PA levels (which were generated as the frequency of a CW Ti:Sapphire ring dye laser for the PA was scanned). The CW dye laser (Coherent 899-29) with a typical power of 1 W and a frequency jitter of ~1 MHz was used to excite those PA levels. The alignment of the PA laser at the center of the overlapped ^{39}K and ^{85}Rb MOTs is checked by optimizing the destruction of the Rb MOT when the laser is tuned to the Rb atomic transition frequency near 12579.00 cm^{-1}.

The calculated branching ratio of singlet and triplet spontaneous emission from those states at 26.5 Å shows that the $2(0^-)$, $3(0^-)$, and $1(2)$ states have pure triplet emission while others have predominant triplet emission [105]. Molecular PA signals were obtained by detecting molecular ion signals generated by RE2PI through the intermediate molecular states using a pulsed dye laser from high vibrational levels of the $X\,^1\Sigma^+$ and $a\,^3\Sigma^+$ states (formed by decay from the PA levels whenever the PA laser frequencies are resonant with the PA levels).

To study excited singlet ($A\,^1\Sigma^+$, $3\,^1\Sigma^+$, and $1\,^1\Pi$) and triplet states ($2\,^3\Sigma^+$ and $b\,^3\Pi$) [106, 107], specific rotational levels of the $3(0^-)$ and $3\,(0^+)$ states, respectively, are chosen. This $3(0^-)$ state decays exclusively to the triplet $a\,^3\Sigma^+$ state and has negligible hyperfine structure. The $3\,(0^+)$ state decays to both the singlet $X\,^1\Sigma^+$ and triplet $a\,^3\Sigma^+$ states. The spectra of the excited singlet ($A\,^1\Sigma^+$, $3\,^1\Sigma^+$, and $1\,^1\Pi$) and triplet ($2\,^3\Sigma^+$ and $b\,^3\Pi$) states were obtained whenever the pulsed laser frequencies were resonant with the vibrational levels of those intermediate singlet and triplet states from high vibrational levels of the $X\,^1\Sigma^+$ and $a\,^3\Sigma^+$ states populated by decay of the PA levels. The pulsed laser beam with a 0.2 cm^{-1} linewidth, a ~5 ns pulse width, a 10 Hz repetition rate, and a few mJ pulse energy was used to obtain the excited singlet and triplet states with a state-selective RE2PI detection method. This pulsed laser was pumped by a frequency-doubled Nd:YAG laser. The scanned spectra were calibrated using a Burleigh wavemeter.

The generated atomic and molecular ions such as Rb$^+$, Rb$_2{}^+$, and K$^+$, and KRb$^+$ produced by this pulsed laser were accelerated into a Channeltron TOF mass detector with low resolution. The KRb$^+$ ion signals produced by this pulsed laser were gated at the KRb$^+$ ion mass, separated from the masses of Rb$^+$, Rb$_2{}^+$, and K$^+$ ions by their times of flight, and integrated using a box car. The resulting KRb$^+$ ion signal as a function of the pulsed laser frequency was recorded using a signal recording software in a personal computer after analog–digital conversion.

References

[1] Smalley R E, Ramakrishna B L, Levy D H and Wharton L 1974 Laser spectroscopy of supersonic molecular beams: application to the NO$_2$ spectrum *J. Chem. Phys.* **61** 4363

[2] Wu C-Y R, Crooks J B, Yang S C, Way K R and Stwalley W C 1973 Li/Li$_2$ supersonic nozzle beam *Rev. Sci. Instrum.* **49** 380

[3] Sinha M P, Schultz A and Zare R N 1973 Internal state distribution of alkali dimers in supersonic nozzle beams *J. Chem. Phys.* **58** 549

[4] Lee Y T, Gordon R J and Herschbach D R 1971 Molecular beam kinetics: reactions of H and D atoms with diatomic molecules *J. Chem. Phys.* **54** 2410

[5] Hsu D S Y, McClelland G M and Herschbach D R 1974 Molecular beam kinetics: angle-angular momentum correlation in reactive scattering *J. Chem. Phys.* **61** 4927

[6] Gordon R J, Lee Y T and Herschbach D R 1971 Supersonic molecular beams of alkali dimers *J. Chem. Phys.* **54** 2393

[7] Kistiakowsky G B and Slichter W P 1951 A high intensity source for the molecular beam. Part II. Experimental *Rev. Sci. Instrum.* **22** 333

[8] Zacharias J R 1942 The nuclear spin and magnetic moment *Phys. Rev* **61** 270

[9] Rousseau S, Allouche A R and Aubert-Frécon M 2000 Theoretical study of the electronic structure of the KRb molecule *J. Mol. Spectrosc.* **203** 235

[10] Foster P J, Leckenby R E and Robbins E J 1969 The ionization potentials of clustered alkali metal atoms *J. Phys. B: At. Mol. Phys.* **2** 478

[11] Mathur B P, Rothe E W, Reck G P and Lightman A J 1981 Two-photon ionization of Li_2: isotopic separation and determination of IP(Li_2) and $D_e(Li^+_2)$ *Chem. Phys. Lett.* **56** 336

[12] Eisel D and Demtröder W 1982 Accurate ionization potential of Li_2 from resonant two-photon ionization *Chem. Phys. Lett.* **88** 481

[13] Engelke F and Hage H 1983 Direct observation of the lowest b $^3\Pi_u$ state of the $^6Li_2$ molecule *Chem. Phys. Lett.* **103** 98

[14] Rubahn H-G and Toennies J P 1988 A molecular beam study of the potential anisotropy of laser vibrationally excited Li_2 ($v = 0$, 20) scattered from Kr *J. Chem. Phys.* **89** 287

[15] Schwarz M, Duchowicz R, Demtröder W and Jungen Ch. 1988 Autoionizing Rydberg states of Li_2: Analysis of electronic–rotational interactions *J. Chem. Phys.* **89** 5460

[16] Ishikawa K, Kubo S and Katô H 1991 The Li_2 C $^1\Pi_u$ state studied by a single-frequency ultraviolet laser *J. Chem. Phys.* **95** 8803

[17] Bouloufa N, Cacciani P, Vetter R and Yiannopoulou A 1999 Tunneling through potential barrier of the B $^1\Pi_u$ state of ^7Li-^7Li *J. Chem. Phys.* **111** 1926

[18] Bouloufa N, Cacciani P, Vetter R, Yiannopoulou A, Martin F and Ross A J 2001 A full description of potential curve of the B $^1\Pi_u$ state of the state of $^7Li_2$ *J. Chem. Phys.* **114** 8445

[19] Bouloufa N, Cacciani P, Kokoouline V, Masnou-Seeuws F, Vetter R and Li L 2001 Predissociation induced by ungerade-gerade symmetry breaking in the B $^1\Pi_u$ state of the $^6Li^7Li$ molecule *Phys. Rev. A* **63** 042507

[20] Sinha M P, Caldwell C D and Zare R N 1974 Alignment of molecules in gaseous transport: Alkali dimers in supersonic nozzle beams *J. Chem. Phys.* **61** 491

[21] Herrmann A, Leutwyler S, Schumacher E and Wöste L 1977 Multiphoton ionization: mass selective laser-spectroscopy of Na_2 and K_2 in molecular beams *Chem. Phys. Lett.* **52** 418

[22] Mathur B P, Rothe E W and Reck G P 1978 Two-photon ionization of Na_2 by an Ar^+ laser *J. Chem. Phys.* **68** 2518

[23] Herrmann A, Schumacher E and Wöste L 1978 Preparation and photoionization potentials of molecules of sodium, potassium, and mixed atoms *J. Chem. Phys.* **68** 2327

[24] Bergmann K, Engelhard R, Hefter U and Hering P 1979 Molecular beam diagnostics with state selection: Intensity Distribution of a Na/Na_2 supersonic beam *Chem. Phys.* **44** 23

[25] Hofmann M, Leutwyler S and Schulze W 1979 Matrix isolation/aggregation of sodium atoms and molecules formed in a supersonic nozzle beam *Chem. Phys.* **40** 145

[26] Herrmann A, Leutwyler S, Wöste L and Schumacher E 1979 Molecular spectroscopy by photodeflection of Na_2 in a supersonic nozzle beam *Chem. Phys. Lett.* **62** 444

[27] Rothe E W, Krause U and Duren R 1980 Observation of polarization of atomic fluorescence excited by laser induced dissociation *Chem. Phys. Lett.* **72** 100

[28] Bergmann K and Gottwald E 1981 Effect of optical pumping in two step photoassociation of Na_2 in molecular beams *Chem. Phys. Lett.* **78** 515

[29] Kompitsas M, Kolwas K and Weber H G 1981 Relaxation in a Na/Na_2 nozzle expansion *Chem. Phys.* **55** 221

[30] Leutwyler S, Hofmann M, Harri H-P and Schumacher E 1981 The adiabatic ionization potentials of the alkali dimers Na_2, NaK and K_2 *Chem. Phys. Lett.* **77** 257

[31] Engelke F, Hage H and Caldwell C D 1982 Na_2 $b\ {}^3\Pi_u - X\ {}^1\Sigma_g^+$ intercombination bands: Direct observation through high resolution laser fluorescence spectroscopy in a supersonic nozzle beam *Chem. Phys.* **64** 221

[32] Delacrétaz G, Ganière J D, Monot R and Wöste L 1982 Photoionization and fragmentation of alkali metal clusters in supersonic molecular beams *Appl. Phys.* B **29** 55

[33] Atkinson J B, Becker J and Demtröder W 1982 Hyperfine structure of the 625 nm band in the $a\ {}^3\Pi_u \leftarrow X\ {}^1\Sigma_g^+$ transitions of Na_2 *Chem. Phys. Lett.* **87** 128

[34] Atkinson J B, Becker J and Demtröder W 1982 Experimental observation of the $a\ {}^3\Pi_u$ state of Na_2 *Chem. Phys. Lett.* **87** 92

[35] Gole J L, Green G J, Pace S A and Preuss D R 1982 The characterization of supersonic sodium vapor expansions including laser induced atomic fluorescence from trimeric sodium *J. Chem. Phys.* **76** 2247

[36] Jones P L, Gaubatz U, Hefter U, Bergmann K and Wellegehausen B 1983 Optically pumped sodium dimer supersonic beam laser *Appl. Phys. Lett.* **42** 222

[37] Delacrétaz G and Wöste L 1985 Two-photon ionization spectroscopy of the (2) ${}^1\Sigma_u^+$ double-minimum state of Na_2 *Chem. Phys. Lett.* **120** 342

[38] Gerber G and Möller R 1985 Optical-optical double-resonance spectroscopy of high vibrational levels of the Na_2 $A\ {}^1\Sigma_u^+$ state in a molecular beam *Chem. Phys. Lett.* **113** 546

[39] Goy P, Bordas M C, Broyer M, Labastie P and Tribollet B 1985 Microwave transitions between molecular Rydberg states *Chem. Phys. Lett.* **120** 1

[40] Haugstätter R, Goerke A and Hertel I V 1988 Case studies in multiphoton ionisation and dissociation of Na_2 *Z. Phys.* D **9** 153

[41] Bordas C, Labastie P, Chevaleyre J and Broyer M 1989 MQDT analysis of rovibrational interactions and autoionization in Na_2 Rydberg states *Chem. Phys.* **129** 21

[42] Bordas C, Broyer M and Vialle J L 1990 Spectroscopy of the 1 ${}^2\Pi_u$ state of Na_2^+ *J. Chem. Phys.* **92** 4030

[43] Kumar S V K, Ziegler G, Korsch H J and Bergmann K 1991 Inelastic transitions in vibrationally excited Na_2 induced by intermediate-energy-electron impact *Phys. Rev.* **44** 268

[44] Richter H, Knöckel H and Tiemann E 1991 The potential of the Na_2 $B\ {}^1\Pi_u$ state *Chem. Phys.* **157** 217

[45] Knöckel H, Johr T, Richter H and Tiemann E 1991 The influence of the spin-orbit and the hyperfine interaction on the asymptotic behaviour of the $A\ {}^1\Sigma_u^+$ state of Na_2 *Chem. Phys.* **152** 399

[46] Zalicki P, Billy N, Gouedard G and Vigue J 1993 Terminal rovibrational distribution of Na_2 in a sodium supersonic beam *J. Chem. Phys.* **99** 6436

[47] Färbert A, Koch J, Platz T and Demtröder W 1994 Vibrationally resolved resonant two-photon ionization spectroscopy of the 1 ${}^3\Sigma_g^+(b) \rightarrow 1\ {}^3\Sigma_u^+(X)$ transition of Na_2 *Chem. Phys. Lett.* **223** 546

[48] Elbs M, Knöckel H, Laue T, Samuelis C and Tiemann E 1999 Observation of the last bound levels near the Na_2 ground-state asymptote *Phys. Rev.* A **59** 3665

[49] Samuelis C, Falke S, Laue T, Pellegrini P, Dulieu O, Knöckel H and Tiemann E 2003 Optical manipulation of long-range interactions at the 3s+3p asymptote of Na_2 *Eur. Phys. J.* D **26** 307

[50] Laue T, Pellegrini P, Dulieu O, Samuelis C, Knöckel H and Tiemann E 2003 Observation of the long-range potential well of the $(6)^1\Sigma_g^+$ (3s+5s) state of Na_2 *Eur. Phys. J.* D **26** 173

[51] Ghazy R, Hamada I M, Demtröder W, El-Kashef H and Hassan G E 2004 Sub-Doppler laser spectroscopy of Na_2 in a cold molecular beam *Egypt. J. Sol.* **27** 77

[52] Garcia-Fernandez R, Ekers A, Klavins J, Yatsenko L P, Bezuglov N N, Shore B W and Bergmann K 2005 Autler-Townes effect in a sodium molecular-ladder scheme *Phys. Rev.* A **71** 023401

[53] Leutwyler S, Hermann A, Wöste L and Schumacher E 1980 Isotope selective two-step photoionization study of K_2 *Chem. Phys.* B **48** 253

[54] Meiwes K H and Engelke F 1982 Predissociation of K_2: molecular beam-laser-induced fluorescence spectroscopy of the $C\,^1\Pi_u - X\,^1\Sigma_g^+$ band system *Chem. Phys. Lett.* **85** 409

[55] Heinze J, Kowalczyk P and Engelke F 1988 Quasibound levels and shape resonances of $^{39}K_2$ $(B\,^1\Pi_u)$ crossed laser-molecular beam studies and analytical interpretation *J. Chem. Phys.* **89** 3428

[56] Kowalczyk P, Schühle U and Engelke F 1989 Rydberg states of the K_2 molecule studied by laser spectroscopy in a supersonic beam *Z. Phys.* D **13** 231

[57] Joo D, Yoon Y, Lee Y, Baek S and Kim B 2000 New electric quadrupole transitions of K_2 observed in a pulsed molecular beam: The $1\,^1\Delta_g - X\,^1\Sigma_g^+$ bands near 500 nm *J. Chem. Phys. (Communication)* **113** 2945

[58] Lisdat C, Knöckel H and Tiemann E 2000 First observation of hyperfine structure in K_2 *J. Mol. Spectrosc.* **199** 81

[59] Falke S, Sherstov I, Tiemann E and Lisdat C 2006 The $A\,^1\Sigma_u^+$ state of K_2 up to the dissociation limit *J. Chem. Phys.* **125** 224303

[60] Sherstov I, Liu S, Lisdat C, Schnatz H, Jung S, Knöckel H and Tiemann E 2007 Frequency measurements in the $b\,^3\Pi_{0u}^+ - X\,^1\Sigma_g^+$ system of K_2 *Eur. Phys. J.* D **41** 485

[61] Falke S, Knöckel H, Friebe J, Riedmann M and Tiemann E 2008 Potassium ground-state scattering parameters and Born-Oppenheimer potentials from molecular spectroscopy *Phys. Rev.* A **78** 012503

[62] Liu S, Sherstov I, Lisdat C, Knöckel H and Tiemann E 2010 Ramsey-Bordé interferometer and embedded Ramsey interferometer with molecular matter waves of $^{39}K_2$ *Eur. Phys. J.* D **58** 369

[63] Lee Y, Yoon Y, Baek S J, Joo D-L, Ryu J and Kim B 2000 Direct observation of the $2\,^3\Pi_u$ state of Rb_2 in a pulsed molecular beam: Rotational branch intensity anomalies in the $2\,^3\Pi_u(1_u) - X\,^1\Sigma_g^+(0_g^+)$ bands *J. Chem. Phys.* **113** 2116

[64] Breford E J and Engelke F 1980 Laser-induced fluorescence in supersonic nozzle beams: predissociation in the Rb_2 $C\,^1\Pi_u$ and $D\,^1\Pi_u$ states *Chem. Phys. Lett.* **75** 132

[65] Caldwell C D, Engelke F and Hage H 1980 High resolution spectroscopy in supersonic nozzle beams: The Rb_2 $B\,^1\Pi_u - X\,^1\Sigma_g^+$ band system *Chem. Phys.* **54** 21

[66] Yoon Y, Lee Y, Lee S and Kim B 2002 Electric quadrupole transitions of Rb_2 observed in a pulsed molecular beam: The $1\,^1\Delta_g - X\,^1\Sigma_g^+$ bands near 540 nm *J. Chem. Phys.* **116** 6660

[67] Lee Y, Lee S and Kim B 2007 Mass-resolved resonance enhanced ionization study of complicated excited electronic states of Rb_2 near 430 nm and their predissociation dynamics *J. Phys. Chem.* A **111** 11750

[68] Lee Y, Lee S and Kim B 2008 Spin-forbidden transitions in the vicinity of the $2\,^1\Pi_u \leftarrow X\,^1\Sigma_g^+$ band system of Rb_2 *J. Phys. Chem.* A **112** 6893

[69] Höning G, Czajkowski M, Stock M and Demtröder W 1979 High resolution laser spectroscopy of Cs_2: I. ground state constants and potential curve *J. Chem. Phys.* **71** 2138

[70] Katô H and Yoshihara K 1979 Laser induced fluorescence, energy transfer and dissociation of Cs_2 *J. Chem. Phys.* **71** 1585

[71] Kim B and Yoshihara K 1993 Determination of adiabatic ionization potentials of Cs_2 and Cs_3 in a very cold molecular beam using time-of-flight mass spectroscopy *Chem. Phys. Lett.* **202** 437

[72] Kim B and Yoshihara K 1993 Triplet-triplet transition of Cs_2 studied by multiphoton ionization spectroscopy in a very cold pulsed molecular beam *Chem. Phys. Lett.* **204** 407

[73] Kim B and Yoshihara K 1993 The 480 nm system of Cs_2 studied in a very cold molecular beam: Direct observation of a new E'' and the ion-pair states *J. Chem. Phys.* **98** 5990

[74] Kim B 1993 Direct observation of the (2) $^3\Pi_u$ state of Cs_2 by resonance enhanced two photon ionization spectroscopy in a very cold molecular beam *J. Chem. Phys.* **99** 5677

[75] Kim B, Yoshihara K and Lee S Y 1994 Complex resonances in the predissociation of Cs_2 *Phys. Rev. Lett.* **73** 424

[76] Lee Y 2002 Excited-state spectroscopy of Rb_2 and KRb by resonance enhanced two photon ionization in a supersonic molecular beam *PhD Thesis* (KAIST, Korea)

[77] Kimura Y, Lefebvre-Brion H, Kasahara S, Katô H, Baba M and Lefebvre R 2000 Interference effects in the predissociation of the Cs_2 $C\ ^1\Pi_u$ and (2) $^3\Pi_u$ states through the dissociative (2) $^1\Sigma_u^+$ state *J. Chem. Phys.* **113** 8637

[78] Bouloufa N, Cacciani P, Vetter R and Yiannopoulou A 2000 Sub-Doppler spectroscopy of the LiH molecule: The $A - X$ system *J. Mol. Spectrosc.* **202** 37

[79] Freeman R R, Jacobson A R, Johnson D W and Ramsey N F 1975 The molecular Zeeman and hyperfine spectra of LiH and LiD by molecular beam high resolution electric resonance *J. Chem. Phys.* **63** 2597

[80] Dagdigian P J 1976 Detection of LiH and NaH molecular beams by laser fluorescence and measurement of radiative lifetimes of the $A\ ^1\Sigma^+$ state *J. Chem. Phys.* **64** 2609

[81] Dagdigian P J and Wharton L 1972 Molecular beam electric deflection and resonance spectroscopy of the heteronuclear alkali dimers: $^{39}K^7Li$, Rb^7Li, $^{39}K^{23}Na$, $Rb^{23}Na$, and $^{133}Cs^{23}Na$ *J. Chem. Phys.* **57** 1487

[82] Pesl F P, Lutz S and Bergmann K 2000 Improved molecular constants for the $X\ ^1\Sigma^+$ and $A\ ^1\Sigma^+$ states of NaH *Eur. Phys. J.* D **10** 247

[83] Breford E J and Engelke F 1978 Laser-induced molecular fluorescence in supersonic nozzle beams: Applications to the NaK $D\ ^1\Pi - X\ ^1\Sigma^+$ and $D\ ^1\Pi - a\ ^3\Sigma^+$ system *Chem. Phys. Lett.* **53** 282

[84] Gerdes A, Dulieu O, Knöckel H and Tiemann E 2011 Stark effect measurements on the NaK molecule *Eur. Phys. J.* D **65** 105

[85] Lee Y, Yun C, Yoon Y, Kim T and Kim B 2001 The 530 nm system of KRb observed in a pulsed molecular beam: New electric quadrupole transitions (1 $^1\Delta - X\ ^1\Sigma^+$) *J. Chem. Phys.* **115** 7413

[86] Lee Y, Yoon Y, Kim B, Li L and Lee S 2004 Observation of the 3 $^3\Sigma^+ - X\ ^1\Sigma^+$ transition of KRb by resonance enhanced two-photon ionization in a pulsed molecular beam: Hyperfine structures of $^{39}K^{85}Rb$ and $^{39}K^{87}Rb$ isotopomers *J. Chem. Phys.* **120** 6551

[87] Lee Y, Yoon Y, Muhammad A, Kim J T, Lee S and Kim B 2010 The 480 nm system of KRb: The 1 $^3\Delta_1$, 4 $^1\Sigma^+$ and 5 $^1\Sigma^+$ states *J. Phys. Chem.* A **114** 7742

[88] Lee Y, Yoon Y, Kim J T, Lee S and Kim B 2011 Unravelling complex spectra of a simple molecule: REMPI study of 420 nm system of KRb *Chem. Phys. Chem.* **12** 2018

[89] Kim B and Yoshihara K 1993 $^3\Delta - {}^1\Sigma^+$ transition of RbCs observed in a very cold molecular bea *Chem. Phys. Lett.* **212** 271

[90] Kim B and Yoshihara K 1994 Resonance enhanced two photon ionization spectroscopy of RbCs in a very cold molecular beam *J. Chem. Phys.* **100** 1849

[91] Yoon Y, Lee Y, Kim T, Ahn J S, Jung Y, Kim B and Lee S 2001 High resolution resonance enhanced two photon ionization spectroscopy of RbCs in a cold molecular beam *J. Chem. Phys.* **114** 8926

[92] Lee Y, Yoon Y, Lee S, Kim J T and Kim B 2008 Parallel and coupled perpendicular transitions of RbCs 640 nm system: mass-resolved resonance enhanced two-photon ionization in a cold molecular beam *J. Phys. Chem.* A **112** 7214

[93] Lee Y, Yoon Y, Lee S and Kim B 2009 500 nm system of RbCs: Assignments and intensity anomalies *J. Phys. Chem.* A **113** 12187

[94] Byer R L and Duncan M D 1981 A 100 msec, reliable, 10 Hz pulsed supersonic molecular beam source *J. Chem. Phys.* **74** 2174

[95] Otis C E and Johnson P M 1970 A simple pulsed valve for use in supersonic nozzle experiments *Rev. Sci. Instrum.* **51** 1128

[96] Skinner G T 1970 A method of detecting viscous effects in a supersonic nozzle beam *Rev. Sci. Instrum.* **41** 1839

[97] Gentry W R and Giese C F 1978 Ten-microsecond pulsed molecular beam source and a fast ionization detector *Rev. Sci. Instrum.* **49** 595

[98] Teshima K and Yasunaga Y 1983 Characteristics of pulsed molecular beams from an electromagnetic valve *Japan J. Appl. Phys.* **22** 1

[99] Auerbach A and McDiarmid R 1980 Modified pulsed valve for supersonic jet applications *Rev. Sci. Instrum.* **51** 1273

[100] Duncan M D, Byer R L and Osterlin P 1981 Pulsed supersonic molecular-beam coherent anti-Stokes Raman spectroscopy of C_2H_2 *Opt. Lett.* **6** 90

[101] Izawa M, Kita S, Takahashi K and Inouye H 1983 A supersonic molecular beam apparatus and characteristics of the beams *Oyo Buturi* **52** 155

[102] Ketterle W, Davis K B, Joffe M A, Martin A and Pritchard D E 1993 High densities of cold atoms in a dark spontaneous-force optical trap *Phys. Rev. Lett.* **70** 2253

[103] Wang H and Stwalley W C 1998 Ultracold photoassociative spectroscopy of heteronuclear alkali-metal diatomic molecules *J. Chem. Phys.* **108** 5767

[104] Wang D, Qi J, Stone M F, Nikolayeva O, Hattaway B, Gensemer S D, Wang H, Zemke W T, Gould P L, Eyler E E and Stwalley W C 2004 The photoassociative spectroscopy, photoassociative molecule formation, and trapping of ultracold $^{39}K^{85}Rb$ *Eur. Phys. J.* D **31** 165

[105] Wang D, Qi J, Stone M F, Nikolayeva O, Hattaway B, Gensemer S D, Wang H, Zemke W T, Gould P L, Eyler E E and Stwalley W C 2004 Photoassociative production and trapping of ultracold KRb molecules *Phys. Rev. Lett.* **93** 243005

[106] Kim J T, Lee Y, Kim B, Wang D, Stwalley W C, Gould P L and Eyler E E 2011 Spectroscopic analysis of the coupled 1 $^1\Pi$, 2 $^3\Sigma^+$ ($\Omega = 0^-$, 1), and b $^3\Pi$ ($\Omega = 0^\pm$, 1, 2) states of the KRb molecule using both ultracold molecule and molecular beam experiments *Phys. Chem. Chem. Phys.* **13** 18755

[107] Kim J T, Lee Y, Kim B, Wang D, Gould P, Eyler E and Stwalley W 2012 Spectroscopic investigation of the A and 3 $^1\Sigma^+$ states of $^{39}K^{85}Rb$ *J. Chem. Phys.* **137** 244301

IOP Concise Physics

Analysis of the Alkali Metal Diatomic Spectra
Using molecular beams and ultracold molecules
Jin-Tae Kim, Bongsoo Kim and William C Stwalley

Chapter 3

Spectroscopy of the KRb molecule

3.1 Previous KRb spectroscopy

In this section we summarize experimental spectroscopic investigations of KRb. Since structureless bands with a peak at 495.9 nm, in the case of the KRb molecule [1], were first observed in the vapor state of the alkali metals, there have been many spectroscopic studies using KRb molecules formed in a heat-pipe oven [2–9], in a MB machine [10–13], in PA [14, 15] of ultracold atoms, in UMs [16–21], and in our combination method [22–25] between MB and UM.

Conventionally, spectroscopy of alkali metal dimers has been carried out in high temperature ovens, e.g. 'heat-pipe ovens' where the optical windows are protected from metal deposition by a layer of cooled buffer gas. Such samples, with many thousands of levels populated, provide spectra which are difficult to analyze unless multiple resonance laser techniques are used. Moreover, for mixtures of two different atoms, two homonuclear species are present in addition to the one heteronuclear species; the spectra of the three species usually overlap. This method prepares molecules near the equilibrium point of the ground state potential well so that it is not easy to access the long-range levels, although there are several exceptions if one has access to very sensitive observation methods. The molecules in the heat-pipe oven are in a thermal Boltzmann population distribution, where many vibrational levels of the ground state are populated. Relatively few molecules are in low rotational levels because the $(2J + 1)$ degeneracy favors higher levels.

Ross *et al* [5] first reported high-resolution spectroscopic information on several bands between the $X\,^1\Sigma^+$ and $A\,^1\Sigma^+$ states. Amiot and Verges [4] constructed PECs of the ground state for low vibrational levels. Pashov *et al* [3] determined a precise dissociation energy $D_e = 4217.815(10)\ \mathrm{cm}^{-1}$ and constructed the PEC of the ground state of the $^{40}\mathrm{K}^{87}\mathrm{Rb}$ molecule up to 14.8 Å combining FTIR spectra (decaying from specific rovibrational levels of the $A\,^1\Sigma^+$ and $1\,^1\Pi$ states) and Feshbach resonances.

Relatively few vibrational levels (with high J') of the $A\,^1\Sigma^+$ state were observed [5]. Lower vibrational levels of the $3\,^1\Sigma^+$ state with ion-pair character at intermediate internuclear distance were observed by the fluorescence transition $3\ ^1\Pi \to 3\ ^1\Sigma^+$ after excitation by Ar^+ laser lines instead of by direct excitation to the $3\ ^1\Sigma^+$ state from the ground state [6].

Kasahara *et al* [8] extensively investigated the $1\ ^1\Pi$ state up to near the dissociation limit and also the $2\ ^1\Pi$ state using polarization labeling spectroscopy. In a subsequent report for the $2\ ^1\Pi$ state, Amiot *et al* [9] corrected by six quanta the vibrational numbering of Kasahara *et al* [8] as well as the T_e value [9]. Low vibrational levels up to the $v' = 15$ vibrational level of the $3\ ^1\Pi$ state observed from excitation spectra between the $X\ ^1\Sigma^+$ and $3\ ^1\Pi$ states were analyzed [6].

Pashov *et al* [3] observed high vibrational levels of the $a\ ^3\Sigma^+$ ($\Omega = 1$) state in the internuclear distance range where Hund's case (c) case can be applied instead of Hund's case (a) and an accurate potential curve of the $a\ ^3\Sigma^+$ state with a dissociation energy of $249.031(10)\ cm^{-1}$ was constructed.

Lower vibrational levels of the $2\ ^3\Sigma^+$ state were observed by the fluorescence transition $3\ ^1\Pi \to 2\ ^3\Sigma^+$ after excitation of the molecules by an Ar^+ laser. Okada *et al* [7] observed four vibrational levels of the $2\ ^3\Sigma^+$ state through perturbations using the technique of Doppler-free optical-optical double resonance polarization spectroscopy. Also they observed two vibrational levels of the $b\ ^3\Pi$ state.

The supersonic MB beam method prepares molecules that are internally cold, although translational velocities are still too high to trap the molecules. The rotational level with maximum population, J''_{max} (typically ~ 5), is much lower than those in a heat-pipe oven (typically > 50). Since 2000, KRb has been investigated by RE2PI in a very cold supersonic MB, where molecules are populated in low rotational levels and the lowest vibrational levels.

Lee *et al* [12] observed the $4\ ^1\Sigma^+$, $5\ ^1\Sigma^+$ and $1\ ^3\Delta_1$ states by using mass-resolved RE2PI in a cold MB. Hyperfine structures of the $3\ ^3\Sigma^+$ state of $^{39}K^{85}Rb$ and $^{39}K^{87}Rb$ [11] were analyzed. However, they could not observe the $4\ ^3\Sigma^+$ and $3\ ^3\Pi$ states (with large Franck–Condon factors (FCFs) between the $X\ ^1\Sigma^+$ and the $4\ ^3\Sigma^+$ and $3\ ^3\Pi$ states in this energy region), because those states do not have significant mixings with singlet states. This method has proven its strength in the detection of the very weak electric quadrupole transitions of $1\ ^3\Delta_1 - X\ ^1\Sigma^+$, the 530 nm system of KRb [10].

Over the years we have investigated these states by using PA [14, 15], UM [16–21], MB [10–13], and combinations of UM and MB [22–25], which prepare basic information for global deperturbation of complex spectra perturbed between multiple electronic states. While the MB technique provides information primarily at short range near the equilibrium distance of the ground state, the others provide information primarily at long range, but with some overlap at intermediate range.

High resolution excitation spectra with only a few quanta of rotational angular momentum ($J' < 5$) of KRb photoassociated via photon absorption from colliding ultracold atoms were investigated extensively by the UConn group [14, 15]. Rovibrational levels of the $2(0^+)$, $2(0^-)$, $2(1)$, $3(0^+)$, $3(0^-)$, $3(1)$, $1(2)$, and $4(1)$ states below the $K\ (4\ S) + Rb\ (5\ P_{1/2})$ asymptote have been analyzed. The observed vibrational levels of the $4(1)$ state match the energies of $v' = 61$–63 vibrational levels of the $1\ ^1\Pi$ state at large internuclear distances as observed by Kasahara *et al* [8]. Later, Stwalley *et al* [26] identified the $v' = 60$ level of the 1 $^1\Pi$ state resonantly coupled with the $v' = 17$ level of the 2 $^1\Pi$ state.

The UM method, starting from ultracold atoms with small kinetic energies, provides a way to prepare molecules at long range and in low rotational levels. We have extensively studied the spectra of $^{39}K^{85}Rb$ UMs formed using previously observed PA levels.

According to Beuc *et al* [27] the electric TDMs between the $a\,^3\Sigma^+$ and $X\,^1\Sigma^+$ states and the 3 $^1\Sigma^+$ state are very small within the corresponding Franck–Condon (FC) windows. Nevertheless, vibrational levels of the 3 $^1\Sigma^+$ state excited from high vibrational levels of the ground $X\,^1\Sigma^+$ state, formed by decay from excited photo-associated molecules, were observed using high resolution depletion spectroscopy by the UConn group [17]. This same depletion technique was also used to obtain the binding energy of the $v'' = 87$, $J'' = 0$ level of the $X\,^1\Sigma^+$ state. Combined with the results of Amiot and Verges [4], this implied a ground state dissociation energy $D_e = 4217.822(3)\,cm^{-1}$, close to the Pashov *et al* [3] value $4217.815(10)\,cm^{-1}$, but well outside the mutual uncertainties. However, Tiemann [28] has criticized the Amiot *et al* results for $v'' = 87$, suggesting the Pashov *et al* D_e value [3] should be used until the small difference is fully resolved. Aikawa *et al* [21] also observed the depletion spectra of $v' = 41$–50 levels of the 3 $^1\Sigma^+$ state of $^{41}K^{87}Rb$ produced via the PA of cold atoms and characterized the transition strength from weakly bound levels of the ground state. Wang *et al* [16] investigated the 4 $^1\Sigma^+$ and 4 $^3\Sigma^+$ states excited from highly excited long-range vibrational levels of the $X\,^1\Sigma^+$ and $a\,^3\Sigma^+$ states differently from those in the MB experiment of Lee *et al* [12]. Vibrational levels $v' = 1$–7 of the 3 $^3\Sigma^+$ state observed in MB by Lee *et al* [11] were extended to $v' = 12$ using an UM experiment [18]. Finally the $v' = 13$ level with a broad linewidth due to tunneling was observed using a UM experiment [19].

The spin–orbit components ($\Omega = 0^+$, 0^-, 1 and 2) of the double-minimum 2 $^3\Pi$ state and a shallow long-range state ($5(0^+)$) at 9.3 Å excited from UMs in high vibrational levels of the $a\,^3\Sigma^+$ state were investigated [18, 29]. Recently Banerjee *et al* [20] investigated low vibrational levels of up to $v' = 12$ of the double-minimum 3 $^3\Pi_\Omega$ state using UM formed by PA.

Finally, a new spectroscopic method using multiplication between MB and UM spectra has been developed to find optimal transfer routes and assignments among complex $A\,^1\Sigma^+$, 3 $^1\Sigma^+$, 1 $^1\Pi$, 2 $^1\Pi$, 2 $^3\Sigma^+$, and $b\,^3\Pi$ spectra [22–24]. KRb is thus now spectroscopically well known among heteronuclear alkali diatomic molecules. However, it is still not complete because of only partial characterization of vibrational levels of specific electronic states, and non-observation of high-lying electronic states. Except for the $a\,^3\Sigma^+$ and $X\,^1\Sigma^+$ states, most of the potentials are incomplete so that hybrid constructions of potential wells of excited states using both experiment and theory is needed to predict unobserved spectra.

From these previous results [1–9, 15–21] we know that only a few of the low vibrational levels of the $A\,^1\Sigma^+$, 3 $^1\Sigma^+$, 1 $^1\Pi$, 2 $^1\Pi$, 2 $^3\Sigma^+$, and $b\,^3\Pi$ states have been reported, except for heat-pipe oven studies of the 1 $^1\Pi$ and 2 $^1\Pi$ states [8] and our UM experiments [15, 16, 22–24] regarding intermediate and high-lying vibrational levels of the $A\,^1\Sigma^+$, $3\,^1\Sigma^+$, $2\,^3\Sigma^+$, and $b\,^3\Pi$ states using our combination analysis of MB and UM spectra.

3.2 MB spectral analysis

Our pulsed MB experiment has proven to be a sensitive experiment which can detect even the very weak electric quadrupole transitions of the $1\,^1\Delta - X\,^1\Sigma^+$ bands of KRb [10]. Alkali molecules with a rotational temperature of less than 5 K are generated by our pulsed MB. The simulation of the rotational contours obtained in the MB experiment shows that only low J'' components (typically $J''_{max} = 5$) are involved. Molecules with a low vibrational temperature of ~1 K for KRb are in vibrational levels $v'' = 0$ (~95%) and $v'' = 1$ (~5%) of the ground state, respectively.

The many mutually perturbing states arising from the K + Rb(5P) and K(4P) + Rb asymptotes, as shown in figure 1.1, make the full assignment of the spectrum difficult due to complex mixtures of Σ and Π character, and of singlet and triplet character, and the fractional parentage of given electronic states can vary rapidly with energy and internuclear distance. We expect strongly perturbed complex spectra in this energy region. However, the following stepwise assignment process made it possible to assign high vibrational levels of the $A\,^1\Sigma^+$, $3\,^1\Sigma^+$, $1\,^1\Pi$, $2\,^3\Sigma^+$, and $b\,^3\Pi$ states in the energy region below the first excited dissociation asymptote.

In MB spectra between $14890\,\mathrm{cm}^{-1}$ and $15850\,\mathrm{cm}^{-1}$ above the minimum of the $X\,^1\Sigma^+$ state we initially found three strong progressions as shown in figure 3.1(a). Then we found one vibrational progression about 100 times weaker than the strong spectra (shown in figure 3.1(b)). In this energy region there are the $A\,^1\Sigma^+$, $3\,^1\Sigma^+$, $1\,^1\Pi$, $2\,^3\Sigma^+$ and $b\,^3\Pi$ states correlating to the K(4S) + Rb($5P_{2/1,3/2}$) asymptotic limits.

We expect the strongest transitions from the strong electric dipole transition moment and the large FCFs between the $X\,^1\Sigma^+$ and $1\,^1\Pi$ states. Molecular constants and term energies of the assigned $1\,^1\Pi$ state observed previously match well with those of theoretical calculation.

In our MB experiment, the initial lower electronic state is the $X\,^1\Sigma^+$ state so that only singlet states can be observed in upper excited states if there are no perturbations between electronic states. Transitions from singlet to triplet states in the MB experiment can be enabled by spin–orbit perturbations that can give singlet character to triplet states (and vice versa) and are usually expected to be much weaker. We observed strong progressions of the $1\,^1\Pi$ state with $\Omega = 1$ which can play a major role in the perturbations. In Hund's case (a), the selection rule for spectroscopic transitions is $\Delta\Omega = 0$, ± 1. For homogeneous perturbations, the selection rule is $\Delta\Omega = 0$, while for the often weaker heterogeneous perturbations, it is $\Delta\Omega = \pm 1$ [30]. Thus the most accessible triplet states for transitions from the initial $X\,^1\Sigma^+$ state in MB experiments are the triplet states with $\Omega = 1$ that can borrow transition intensity from the $1\,^1\Pi$ state. Strong spin-forbidden triplet transitions from the $X\,^1\Sigma^+$ ($v'' = 0$ and $v'' = 1$) levels were observed because of strong spin–orbit mixing between singlet $1\,^1\Pi$ and triplet $2\,^3\Sigma^+$ and $b\,^3\Pi$ states with $\Omega = 1$.

Thus, in addition to singlet states, we initially observed only the $\Omega = 1$ component of triplet states in the MB spectra. In the UM+ and UM− experiments, PA is used

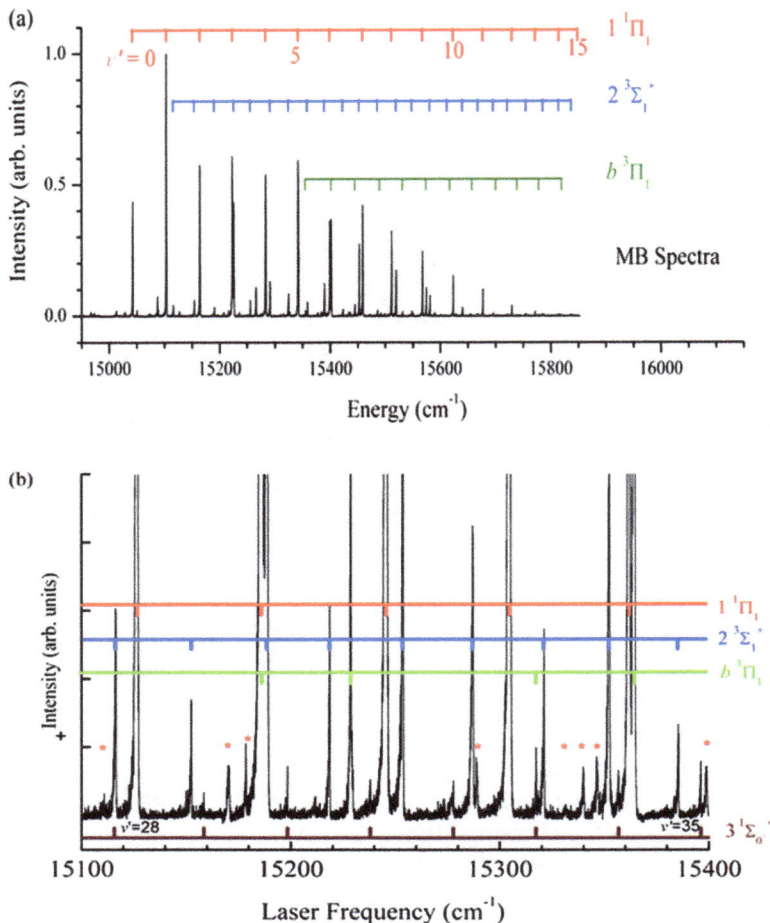

Figure 3.1. (*a*) Complete spectra of $^{39}K^{85}Rb$ from MB experiment between $14\,890\,cm^{-1}$ and $15\,850\,cm^{-1}$, reproduced from [22]. (*b*) Spectra of the $3\,^1\Sigma^+$, $1\,^1\Pi$, $2\,^3\Sigma^+$, and $b\,^3\Pi$ states from our MB experiment. The energy ranges covered are between $15\,100\,cm^{-1}$ and $15\,400\,cm^{-1}$. The observed MB spectra are expanded in signal intensity to show very weak signals corresponding to the $3\,^1\Sigma^+$ state which are not visible in (a). Hot bands from the $v'' = 1$ level of the $X\,^1\Sigma^+$ state and unassigned vibrational levels are marked by the symbols (*) and (+), respectively. The energy is given with respect to the minimum of the ground $X\,^1\Sigma^+$ state potential well (i.e., the spectrum is shifted up by $E(X\,^1\Sigma^+, v'' = 0, J'' = 0)$ [$37.864\,cm^{-1}$] to the excitation laser frequency in cm^{-1}).

to form a specific rovibrational level of the $3(0^+)$ and $3(0^-)$ states, respectively. In the UM− and UM+ spectra, all spin–orbit components of triplet states were observed (2 for $^3\Sigma^+$ and 4 for $^3\Pi$). A simple comparison of excited state energies in the MB spectra with those in the UM+ or UM− spectra allowed a complete assignment.

For example, the $2\,^3\Sigma^+$ state has two components with $\Omega = 0^-$ and 1 while the $b\,^3\Pi$ state has four components with $\Omega = 0^\pm$, 1, and 2. Each $\Omega = 0^-$ component can only be accessed from the 0^- component of the $a\,^3\Sigma^+$ state in the UM− experiment, but it is not accessible in the MB experiment due to the selection rule $+ \leftrightarrow +$, $- \leftrightarrow -$, and $+ \leftarrow/\rightarrow -$ [31].

Figures 3.2(a), (b), and (c) show the UM−, UM+, and MB spectra between $15\,370\ \mathrm{cm}^{-1}$ and $15\,600\ \mathrm{cm}^{-1}$, respectively, which are explained in detail in sections 3.3.1 and 3.3.2. In the UM− spectra of figure 3.2(a), clear Ω doublet structures due to spin–orbit interaction are visible. Energy positions of vibrational levels of the $2\ ^3\Sigma^+$ state with $\Omega = 1$ in the MB spectra match well with those in the UM− spectra so that those were assigned as $\Omega = 1$ and others were assigned as $\Omega = 0^-$. Thus, the $\Omega = 1$ and $\Omega = 0^-$ components of the $2\ ^3\Sigma^+$ state can be completely assigned. Other remaining progressions with moderate signal strengths (except two progressions with very weak signal strengths) also have more than two spin–orbit components in UM spectra and were assigned to the $b\ ^3\Pi$ state with $\Omega = 1$ using the same method as above.

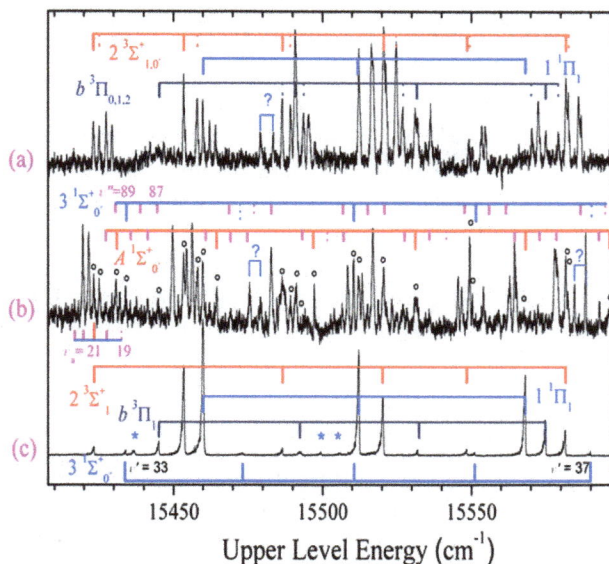

Figure 3.2. (a) Assigned UM− spectra of the $1\ ^1\Pi$, $2\ ^3\Sigma^+$ (with $\Omega = 0$ (red dotted lines) lying above $\Omega = 1$ (red solid lines)), and $b\ ^3\Pi$ (with $\Omega = 0$, 2 (blue dotted lines) lying beside $\Omega = 1$ levels (blue solid lines) excited from high vibrational levels of the $a\ ^3\Sigma^+$ state, populated by decay from a specific level of the $3(0^-)$ state (formed by PA). (b) Assigned UM+ spectra of the $A\ ^1\Sigma^+$, $3\ ^1\Sigma^+$, $1\ ^1\Pi$, $2\ ^3\Sigma^+$, and $b\ ^3\Pi$ levels excited from high vibrational levels of the $X\ ^1\Sigma^+$ and $a\ ^3\Sigma^+$ states formed by decay from a specific level of the $3(0^+)$ state (formed by PA) (c) Assigned MB spectra of the $3\ ^1\Sigma^+$, $1\ ^1\Pi$, $b\ ^3\Pi$, and $2\ ^3\Sigma^+$ levels excited from low vibrational levels ($v'' = 0$ and 1) of the $X\ ^1\Sigma^+$ state. Hot bands from the $v'' = 1$ level of the $X\ ^1\Sigma^+$ state vibrational levels are marked by the symbol (*). All of the assigned electronic levels in panel (b) that match well with assigned vibrational levels in figure 3.2(a) and (c) are marked by the symbol (o). It should be noted that the energy scales in the MB, UM+, and UM− spectra are all the same and are relative to the minimum of the ground $X\ ^1\Sigma^+$ state potential. Thus, each spectrum is shifted up by the energy of the lower level of the excitation (relative to the bottom of the ground-state potential well): the frequency axes of the UM−, UM+, and MB spectra shown in figure 3.2 are shifted up by the energies $E(a\ ^3\Sigma^+, v_a = 21, J_a = 0)$ [$4202.976\ \mathrm{cm}^{-1}$], $E(X, v'' = 89, J'' = 0)$ [$4204.762\ \mathrm{cm}^{-1}$], and $E(X\ ^1\Sigma^+, v'' = 0, J'' = 0)$ [$37.864\ \mathrm{cm}^{-1}$] relative to the excitation laser frequency in cm^{-1}, respectively. Reproduced from [23].

While the FCFs between the $X\,^1\Sigma^+$ ($v''=0$) and $b\,^3\Pi$ (v') states are almost negligible, weak transitions can nevertheless be observed to the most perturbed levels in this energy region for the MB experiment. For the UM+ and UM− experiments, the $b\,^3\Pi$ state bands are much stronger, because both the FCFs and the dipole transition moments are stronger.

We expect that MB transitions to the $3\,^1\Sigma^+$ state with $\Omega=0^+$ will be weak, because of the very small electric dipole transition moment between the $X\,^1\Sigma^+$ and $3\,^1\Sigma^+$ states [27]. Nevertheless we observe a regular and very weak vibrational progression assignable to the $3\,^1\Sigma^+$ (v') $\leftarrow X\,^1\Sigma^+$ (v'') transition in the MB excitation spectra as shown in figure 3.1(b), where the intensity scale has been expanded.

After eliminating assigned bands of the $1\,^1\Pi$, $b\,^3\Pi$, and $2\,^3\Sigma^+$ states, only one regular vibrational progression with very weak signal intensities in the MB experiment remains unassigned. Also observed are 'hot bands', marked by the symbol (*), weak signals from the $v''=1$ level which are separated by 75.38 cm^{-1} from signals of the $v''=0$ level. The vibrational spacing of this progression ($\sim$39 cm^{-1}) is close to that calculated for the $3\,^1\Sigma^+$ state. Thus we assign this progression to vibrational levels between $v'=24$ and $v'=43$ of the $3\,^1\Sigma^+$ state. Most of the observed MB spectra including hot bands are now assigned. Higher vibrational levels (above $v'=43$) are observed using the vibrational progression of the $3\,^1\Sigma^+$ state in the UM+ spectra. We can then further extend the vibrational assignments of that state to higher vibrational levels by using the UM+ spectra from the higher vibrational levels of the $X\,^1\Sigma^+$ and $a\,^3\Sigma^+$ states.

We expect that transitions to the $A\,^1\Sigma^+$ state with $\Omega=0^+$ from the $X\,^1\Sigma^+$ state may be very weak due to very small FCFs. The transitions to the $A\,^1\Sigma^+$ state with $\Omega=0^+$ from the $X\,^1\Sigma^+$ state were not observed in the MB experiment. However, these transitions were observed in the UM+ experiment described in section 3.3.2.

3.3 UM spectral analysis

In the UM experiments we used two different PA levels of different symmetry ($3(0^-)$ and $3(0^+)$ levels) to investigate high vibrational levels of the $A\,^1\Sigma^+$, $3\,^1\Sigma^+$, $1\,^1\Pi$, $2\,^3\Sigma^+$, and $b\,^3\Pi$ states. The UM experiments using the $3(0^-)$ and $3(0^+)$ PA levels are designated as UM− and UM+, respectively. Both of the UM spectra have only vibrational resolution, but only low J' components are involved because only the $J'=1$ levels of the $3(0^+)$ and $3(0^-)$ states are initially excited by PA. These two levels have different symmetries, so their selection rules are different. Radiative decay from the $3(0^-)$ level can occur only to the $a\,^3\Sigma^+$ state because of the symmetry selection rules [31]. However, the $3(0^+)$ level decays to both the $X\,^1\Sigma^+$ and $a\,^3\Sigma^+$ states (although only to the $\Omega=1$ component of the a state). More specifically, in the UM+ spectra, $X\,^1\Sigma^+$ ($v''=87$–90) and $a\,^3\Sigma^+$ ($v_a=19$–23) are populated from the decay of the $3(0^+)$ PA level, while in the UM− spectra, only $a\,^3\Sigma^+$ ($v_a=20$ and 21) are populated. Also the singlet $A\,^1\Sigma^+$ and $3\,^1\Sigma^+$ states were not observed in the UM− spectra although those were observed in the UM+ spectra. However, the triplet states were observed in both UM− and UM+ spectra. The UM experiment provides more complex spectra compared with those from the MB experiment because UM spectra involve several high vibrational levels of $a\ ^3\Sigma^+$ state, and access a much broader range of internuclear separations R, while the MB spectra involve only $v'' = 0$ and $v'' = 1$.

3.3.1 UM− spectral analysis

The decay of the $3(0^-)$ state is consistent with the dipole transition selection rules in Hund's case (c), where an $\Omega = 0^-$ state can emit only to $\Omega = 0^-$ and 1 states and cannot decay into the $\Omega = 0^+$ state due to selection rules $(+ \leftrightarrow +, - \leftrightarrow -)$. Figure 3.2 shows (a) the UM− and (b) UM+ spectra, respectively.

In the UM− spectra, spectral features show a recurring doublet of bands with comparable intensity, with an approximate $4.23\ \text{cm}^{-1}$ spacing, the difference between the $v_a = 20$ and 21 levels of the $a\ ^3\Sigma^+$ state. This occurs because both vibrational levels are populated by the initial PA. For the UM− experiment the initial $a\ ^3\Sigma^+$ state has $\Omega = 0^-$ and 1, so the accessible final states are triplet states with $\Omega = 0^\pm$, 1, and 2 and singlet states with $\Omega = 0^+$ and 1.

There is a major difference in Ω components between singlet and triplet states. There are two components ($\Omega = 0^-$ and 1) for the $2\ ^3\Sigma^+$ state and four components ($\Omega = 0^\pm$, 1, and 2) for the $b\ ^3\Pi$ state, while the singlet $A\ ^1\Sigma^+$ ($\Omega = 0^+$), $3\ ^1\Sigma^+$ ($\Omega = 0^+$), and $1\ ^1\Pi$ ($\Omega = 1$) states do not have multiple spin–orbit components.

A vibrational progression with a doublet structure for the $2\ ^3\Sigma^+$ state was found easily comparing the energy positions of $\Omega = 1$ component of the $2\ ^3\Sigma^+$ and $b\ ^3\Pi$ states with those assigned in the MB experiment. Also the $1\ ^1\Pi$ state is easily sorted out by comparing the energy positions with those assigned in the MB experiment. For the UM− experiment, transition strengths of the $2\ ^3\Sigma^+$ and $b\ ^3\Pi$ states are much stronger, as shown in figure 3.2, because both the FCFs and the dipole transition moments are stronger compared to those in the MB experiment. The $A\ ^1\Sigma^+$ and $3\ ^1\Sigma^+$ states were not observed in the UM− spectra although those were observed in the UM+ spectra. The reasons why observations of those spectra depend on PA levels such as the $3(0^+)$ and $3(0^-)$ levels used for the UM experiments is described by considering Hund's case (c) selection rules and the TDM calculations of Kotochigova $et\ al$ [33] between the upper excited $A\ ^1\Sigma^+$ ($2(0^+)$) state and the three Ω components at the ground-state dissociation limit. Presumably this can also be applied to the $3\ ^1\Sigma^+$ state, but unfortunately there are no corresponding calculations for the $3\ ^1\Sigma^+$ state. This is discussed in detail in section 3.4.4.

3.3.2 UM+ spectral analysis

In the UM+ spectra as shown in figure 3.2(b), $X\ ^1\Sigma^+$ ($v'' = 87$–90) and $a\ ^3\Sigma^+$ ($v_a = 19$–23) are populated from the decay of the $3(0^+)$ PA level, while in the UM− spectra, only $a\ ^3\Sigma^+$ ($v_a = 20$ and 21) levels are populated. Also the $A\ ^1\Sigma^+$ and $3\ ^1\Sigma^+$ states are observed differently: they are absent in the UM− spectra, but present in the UM+ spectra; we discuss the reasons why such cases happen in section 3.4.4. The UM+ spectra from high-lying vibrational levels of the $X\ ^1\Sigma^+$ and $a\ ^3\Sigma^+$ states, including those to the $A\ ^1\Sigma^+$ and $3\ ^1\Sigma^+$ states, are more complicated than the UM− spectra. We have been able to assign the $3\ ^1\Sigma^+$, $1\ ^1\Pi$, $2\ ^3\Sigma^+$, and $b\ ^3\Pi$ states by combining the strong transitions in the

MB and UM− spectra from high-lying vibrational levels of the $a\,{}^3\Sigma^+$ state. In the MB experiment, as explained in the previous section, we also assign weak transitions to lower vibrational levels of the $3\,{}^1\Sigma^+$ state. We then extend the vibrational assignments of that state to higher vibrational levels by using the UM+ spectra, which include excitation from the higher vibrational levels of the $X\,{}^1\Sigma^+$ and $a\,{}^3\Sigma^+$ states.

We assign the $A\,{}^1\Sigma^+$ state to the transitions which still remain unassigned from the MB and UM− spectra, using the UM+ spectra to confirm the identifications by comparing the spectra from the $X\,{}^1\Sigma^+$ and $a\,{}^3\Sigma^+$ vibrational levels. Finally we successfully sort out the $A\,{}^1\Sigma^+$ and $3\,{}^1\Sigma^+$ states among the many observed spectroscopic lines to the $A\,{}^1\Sigma^+$, $3\,{}^1\Sigma^+$, $1\,{}^1\Pi$, $2\,{}^3\Sigma^+$, and $b\,{}^3\Pi$ states by using UM+, UM−, and MB experiments in the strongly perturbed energy region between $14890\ \mathrm{cm}^{-1}$ and $16300\ \mathrm{cm}^{-1}$ above the minimum of the ground $X\,{}^1\Sigma^+$ state. The energies of the assigned $A\,{}^1\Sigma^+$ levels are close to theoretical predictions in supplementary information reported previously [34].

Most lines have been assigned in the strongly perturbed region above $15140\ \mathrm{cm}^{-1}$ as shown in figure 3.1(b). We could not assign some of the observed weak lines below $15140\ \mathrm{cm}^{-1}$. This region includes the avoided crossings of the $\Omega = 0^-$ and 1 potential curves of the $b\,{}^3\Pi$ and $2\,{}^3\Sigma^+$ electronic states at ~6.4 Å, visible in figure 1.1.

Assigned UM+ spectra of the $A\,{}^1\Sigma^+$, $3\,{}^1\Sigma^+$, $1\,{}^1\Pi$, $2\,{}^3\Sigma^+$, and $b\,{}^3\Pi$ states, starting in high vibrational levels of the $X\,{}^1\Sigma^+$ and $a\,{}^3\Sigma^+$ states, are shown in figure 3.2(b). All assigned electronic states which match well with assigned vibrational levels in figures 3.2(a) and (c) are marked by the symbol (o) in figure 3.2(b). The assigned vibrational levels of the initial electronic states include the $v'' = 89$ level of the $X\,{}^1\Sigma^+$ state and $v_a = 21$ level of the $a\,{}^3\Sigma^+$ state. Other (usually weaker) lines are from $v_a = 19$, 20, 22, and 23 of the $a\,{}^3\Sigma^+$ state and from $v'' = 87$, 88, and 90 of the $X\,{}^1\Sigma^+$ state. We note that the energy splitting ($3.61\ \mathrm{cm}^{-1}$) between $v_a = 21$ and $v_a = 22$ of the $a\,{}^3\Sigma^+$ state is only slightly larger than that between $v'' = 89$ and $v'' = 90$ of the $X\,{}^1\Sigma^+$ state. Transitions from the $a\,{}^3\Sigma^+$ state to the excited upper states are weaker than those from the $X\,{}^1\Sigma^+$ state to the excited upper states. Two unassigned transitions marked by the symbol (?) still remain in this energy region. Vibrational levels $v' = 35$–51 of the $3\,{}^1\Sigma^+$ state and $v' = 93$–122 of the $A\,{}^1\Sigma^+$ state have been assigned in the UM+ experiment.

3.4 Spectroscopic results from MB and UM spectroscopy

3.4.1 Term energies of observed electronic states

Vibrational numberings from vibrational progressions of the $A\,{}^1\Sigma^+$, $3\,{}^1\Sigma^+$, $1\,{}^1\Pi$, $2\,{}^3\Sigma^+$, and $b\,{}^3\Pi$ electronic states assigned in previous sections were made by comparing term energies from theoretical calculations [35] with the assigned experimental ones. In the UM experiments, UM+ spectra were excited from vibrational levels in $v'' = 87$–90 of the ground $X\,{}^1\Sigma^+$ state (formed by decay from the $3(0^+)$ PA level), but UM− spectra were excited from only two $v_a = 21$ and 22 vibrational levels of the $a\,{}^3\Sigma^+$ state (formed by decay from the $3(0^-)$ PA level). In the MB experiment, MB spectra were excited from vibrational $v'' = 0$ and 1 levels of the ground $X\,{}^1\Sigma^+$ state.

Thus to obtain $T_{v'}$ values of the electronic states assigned in the UM+ spectra, the laser excitation frequencies were added to the energy $(D_e\ (X\,{}^1\Sigma^+) - \Delta E(X\,{}^1\Sigma^+,$

$v'' = 89$, $J'' = 0$)), where $D_e(X\,^1\Sigma^+)$ and ΔE are the dissociation energy [17] of the $X\,^1\Sigma^+$ state and the binding energies of the energy levels of the relevant states. To obtain $T_{v'}$ values of the electronic states assigned in the UM− spectra, the laser excitation frequencies were added to the energy ($D_e\,(X\,^1\Sigma^+) - \Delta E(a\,^3\Sigma^+$, $v_a = 21$, $J_a = 0$)), where ΔE is the binding energy of the energy levels of the relevant states. To obtain $T_{v'}$ values of the electronic states assigned in the MB spectra, the laser excitation frequencies were added to the energy $E(X, v'' = 0, J'' = 0)$ (37.864 cm^{-1}). The supplementary data tables reported previously [34, 36] show vibrational number assignments and term energies for the $A\,^1\Sigma^+$, $3\,^1\Sigma^+$, $1\,^1\Pi$, $2\,^3\Sigma^+$, and $b\,^3\Pi$ electronic states. Also term energies reported previously by Kasahara $et\ al$ [8] and Okada $et\ al$ [7] are shown [34]. Kasahara $et\ al$ [8] observed rotationally resolved transitions to a few vibrational levels of the $2\,^3\Sigma^+$ and $b\,^3\Pi_1$ states. While Kasahara $et\ al$ [8] provide accurate results with their very high spectral resolution (~0.0004 cm^{-1}), our results match well within our spectral resolution (±0.09 cm^{-1}). They also agree well with those of Okada $et\ al$ [7] and Kasahara $et\ al$ [8].

3.4.2 Comparison between experimental $\Delta G_{v'+1/2}$ and theoretical values

For regular potential curves of the electronic states without perturbations, vibrational level spacings decrease smoothly with increasing v. Change in the vibrational level spacing reflects the deviation from a Morse potential well. The energy limit of the vibrational levels provides the dissociation limit of the electronic state. However, vibrational levels close to the dissociation energy are governed by the long range potential. We compare plots of the experimental vibrational level spacing with those obtained from vibrational energies of $ab\ initio$ PECs of the related electronic states.

The vibrational energy $G(v)$ can be determined from equation (3.1) by using experimental molecular constants [37]

$$G(v) = \sum_{i=1}^{\infty} Y_{i0}\left(v + \frac{1}{2}\right)^i = Y_{10}\left(v + \frac{1}{2}\right) + Y_{20}\left(v + \frac{1}{2}\right)^2 + ...,\qquad(3.1)$$

where Y_{i0} corresponds to vibrational constants among Dunham coefficients (Y_{ij}) fitted by experimentally obtained molecular energies. The vibrational level spacing can be written as $\Delta G_{v+1/2} = G(v+1) - G(v)$.

Here we discuss vibrational spacings $\Delta G_{v'+1/2}$ of the $A\,^1\Sigma^+$, $3\,^1\Sigma^+$, $1\,^1\Pi$, $2\,^3\Sigma^+$, and $b\,^3\Pi$ states. The magnitude of the off-diagonal spin–orbit terms between these states becomes comparable with the vibrational spacings of the mutually perturbed electronic states corresponding to an intermediate Hund's coupling case between (a) and (c). Thus we expect the vibrational spacings for these states to be quite irregular.

The spacings between successive vibrational levels, $\Delta G_{v'+1/2}$, are obtained from the assigned bands of the $1\,^1\Pi$ state from the MB experiment and are shown in figure 3.3(a). The $\Delta G_{v'+1/2}$ values of the $1\,^1\Pi$ state abruptly decrease when the vibrational quantum number is near $v' = 8$ due to the avoided crossing of the $1\,^1\Pi$ and $2\,^1\Pi$ states as shown in figure 1.1. Due to perturbations for levels above $v' = 28$, the $\Delta G_{v'+1/2}$ values observed by Kasahara $et\ al$ [8] fluctuate as shown in figure 3.3(a). Figures 3.3(b) and (c) show $\Delta G_{v'+1/2}$ comparisons for the $2\,^3\Sigma^+$ ($\Omega = 1$) state and

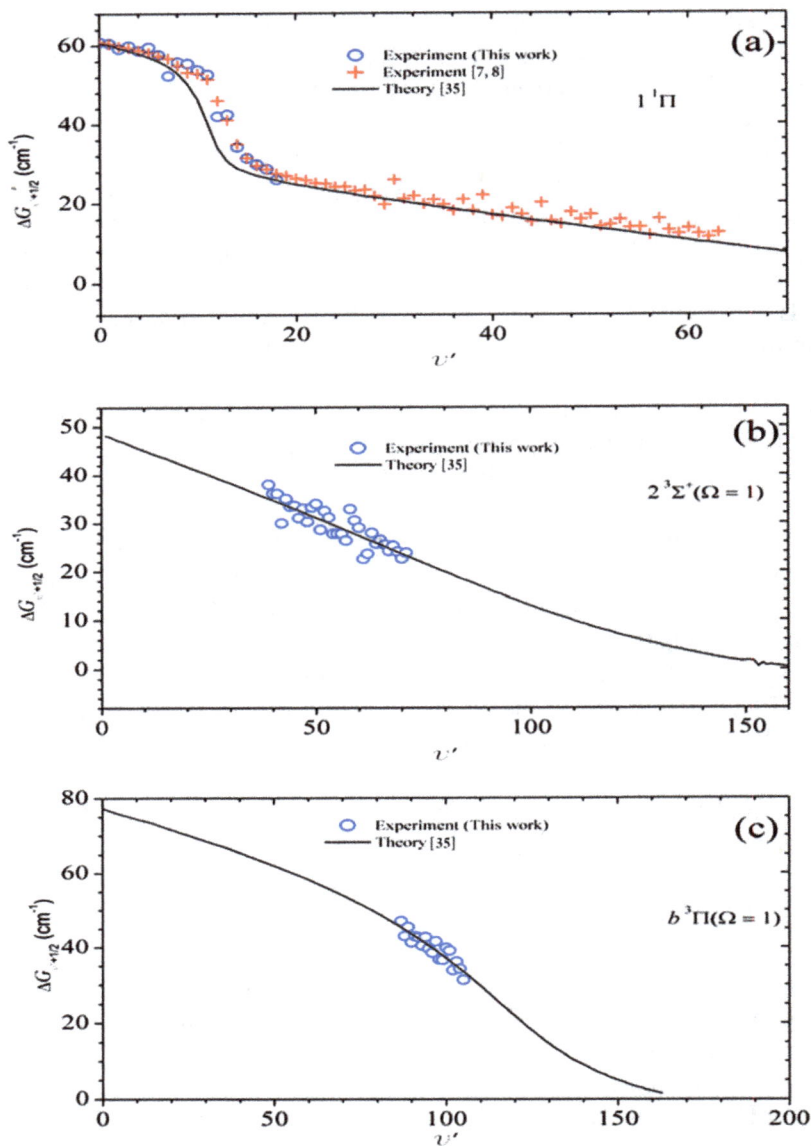

Figure 3.3. (a) $\Delta G_{v'+1/2}$ comparison for the 1 $^1\Pi$ state showing theory based on the PECs of reference [35] (solid line), previous experimental results (+) [7, 8], and results from our **MB** experiment (o). Our experimental vibrational spacings agree well with those of references [7] and [8], with one exception. (b) A similar comparison for the 2 $^3\Sigma^+$ ($\Omega = 1$) state between theory (solid line) and our **MB** experiment (o) shows good overall agreement, but with appreciable scatter. (c) Same for the b $^3\Pi$ ($\Omega = 1$) state, again showing good agreement of overall trends. Reproduced from [22].

$b\,^3\Pi$ ($\Omega = 1$) states between theory (line) and our MB experiment (o). Our experimental vibrational spacings show obvious perturbations, but otherwise follow the trend from those calculated from the theoretical PECs of Rousseau *et al* [35]. This observed progression results in $\Delta G_{v'+1/2} \cong 40\,\mathrm{cm}^{-1}$ for the $b\,^3\Pi$ state in this energy region, as shown in figure 3.3(*c*).

The 2 $^1\Pi$, 2 $^3\Pi$, and 3 $^3\Sigma^+$ states, which lie far above the 3 $^1\Sigma^+$ state in our experimental energy region, also dissociate to the excited K dissociation limits as shown in figure 1.1. However, the $A\,^1\Sigma^+$ state dissociates to the $K(4S) + \mathrm{Rb}(5P_{1/2})$ limit and is perturbed strongly by nearby electronic states such as the 1 $^1\Pi$, 2 $^3\Sigma^+$, and $b\,^3\Pi$ states due to the spin–orbit interaction. Thus we expect the 3 $^1\Sigma^+$ state to be perturbed less strongly by these states, accounting for its more regular vibrational progression compared to the $A\,^1\Sigma^+$ state.

$\Delta G_{v'+1/2}$ values of the $A\,^1\Sigma^+$ and 3 $^1\Sigma^+$ states are shown in figure 3.4 as a function of energy, where zero energy is referred to the minimum of the ground-state potential. The $\Delta G_{v'+1/2}$ values of the $A\,^1\Sigma^+$ state are intermittently observed due to the distribution of FCFs. They are more irregular than those of the 3 $^1\Sigma^+$ state due to the perturbations, and also smaller in the energy region above $15618\,\mathrm{cm}^{-1}$. It is characteristic of ion-pair states [38] (such as the 3 $^1\Sigma^+$ state) for the intervals $\Delta G_{v'+1/2}$ to

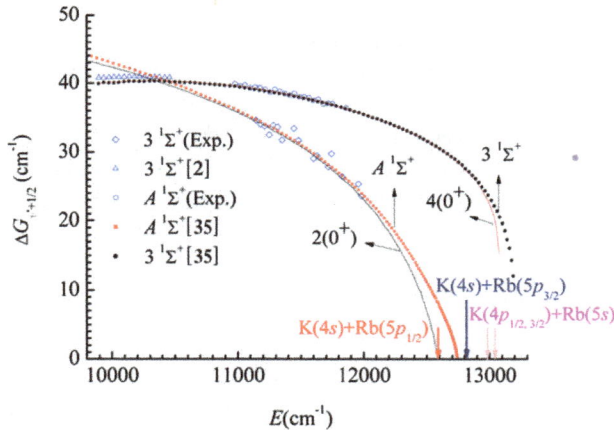

Figure 3.4. Vibrational spacings $\Delta G_{v'+1/2}$ of the $A\,^1\Sigma^+$ and 3 $^1\Sigma^+$ states with an energy scale referred to the minimum of the ground-state potential. Theoretical energies are shifted up by $4,217.822\,\mathrm{cm}^{-1}$, where $4,217.822\,\mathrm{cm}^{-1}$ is the value of the ground-state dissociation energy from reference [17]. Our experimental observations of the $A\,^1\Sigma^+$ and 3 $^1\Sigma^+$ states are marked by symbols ($\diamond$) and ($\circ$), respectively. Theoretical $\Delta G_{v'+1/2}$ values of the $A\,^1\Sigma^+$ and 3 $^1\Sigma^+$ states based on the potential curves of Rousseau *et al* [35] are marked by symbols ($\bullet$) and ($\blacksquare$), respectively. Theoretical $\Delta G_{v'+1/2}$ values of the $2(0^+)$ and $4(0^+)$ states based on the potential curves of Rousseau *et al* [35] are marked by solid lines. Prior experimental values for the 3 $^1\Sigma^+$ state by Amiot *et al* [2] are marked by the symbol ($\triangle$). The $\Delta G_{v'+1/2}$ plot for the 3 $^1\Sigma^+$ state has a maximum in the lower energy region reflecting the ion-pair character of the potential. Note that the 3 $^1\Sigma^+$ state is less perturbed than the $A\,^1\Sigma^+$ state, as indicated by its more regular $\Delta G_{v'+1/2}$ spacings. At the high-v' limit, 3 $^1\Sigma^+$ and $4(0^+)$ states are predicted to have 9 and 6 quasibound levels, respectively. The last levels of the 3 $^1\Sigma^+$ and $4(0^+)$ states are the only ones which predominantly predissociate by tunneling rather than radiate, with very short estimated theoretical lifetimes of $1.0 \times 10^{-10}\,\mathrm{s}$ and $3.9 \times 10^{-12}\,\mathrm{s}$, respectively. Reproduced from [22].

initially increase with energy, then pass through a maximum before decreasing as for a covalent potential. The $3\,^1\Sigma^+$ state dissociates adiabatically to the $K(4P_{1/2}) + Rb(5S)$ limit.

3.4.3 Energy ordering reversal of Ω levels for the $2\,^3\Sigma^+$ state

The component Ω of the total electronic angular momentum along the internuclear axis of electronic states is associated with the components of the electronic orbital angular momentum Λ and the electronic spin Σ along the internuclear axis ($\Omega = |\Lambda + \Sigma|$). Thus the $^3\Pi$ state has four Ω components although the $^3\Sigma^+$ state has two Ω components. The combination method between MB and UM makes it easier to identify Ω components irrespective of the energy order of the Ω components. We found a reversal in energy of the $\Omega = 0^-$ and $\Omega = 1$ components of the $2\,^3\Sigma^+$ state as the energy increased near $15600\ \text{cm}^{-1}$ (see figure 3.5).

Here we compare our experimental observation of Ω reversal in energy positions with obtained eigenvalues of theoretical Hund's case (c) PECs of the $2\,^3\Sigma_\Omega^+$ ($\Omega = 2(0^-)$ and $2(1)$) state using our assigned spectra of the $2\,^3\Sigma^+$ state. Figure 3.5 shows the energy difference between the $2(1)$ and $2(0^-)$ states obtained by using *ab initio* PECs and from the experiment. The *ab initio* energy difference increases initially, reaching a maximum at the avoided crossing point and then decreasing and becoming negative. The zero crossing occurs near $15\,000\ \text{cm}^{-1}$ theoretically and is seen near $15\,624\ \text{cm}^{-1}$ experimentally. At R values beyond the crossing of the PECs, the $\Omega = 1$ state is lower in energy than the $\Omega = 0^-$ state at the same internuclear distance, while inside the crossing the order is reversed. This behavior is unique to the $2\,^3\Sigma^+$ state in our observations. According to previous calculations, the spin–orbit separations between the $\Omega = 0^-$ and $\Omega = 1$ states was $2\ \text{cm}^{-1}$ for the $2\,^3\Sigma^+$ state. The small and variable $2\,^3\Sigma^+$ splitting occurs because a pure $\Lambda = 0$ configuration would exhibit very little

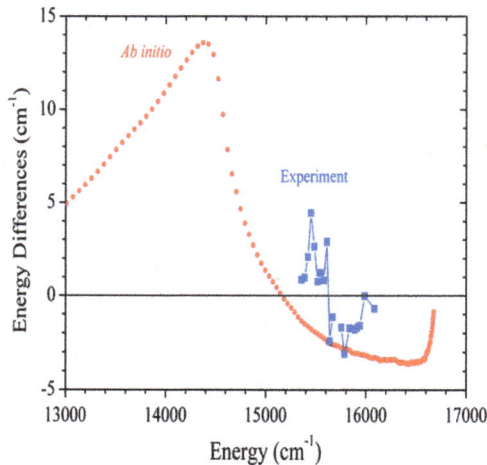

Figure 3.5. The energy difference between the $2(1)$ and $2(0^-)$ states crosses zero near $15\,000\ \text{cm}^{-1}$ for the *ab initio* potentials [35], but near $15\,624\ \text{cm}^{-1}$ for the experiment. Reproduced from [22].

splitting, but the state can acquire $^3\Pi$ character due to mixing with the $b\ ^3\Pi$ state. The Ω reversal in the $2\ ^3\Sigma^+$ state could not have been conclusively demonstrated without using the spectra from both the MB and the UM experiments. By comparing them, it is unambiguous whether the $\Omega = 1$ component or the $\Omega = 0^-$ component has the higher energy. However, in the non-overlapping region above $15\,850\,\mathrm{cm}^{-1}$, this comparison is not yet possible and the ordering can be determined only from continuity and comparison with theory.

3.4.4 Confirmation of $^1\Sigma^+$ states observation using selection rules and TDMs

We observe that the $A\ ^1\Sigma^+$ and $3\ ^1\Sigma^+$ states are observable in the UM+ spectra, but absent from the UM− spectra. This is explained by considering the following Hund's case (c) selection rules and TDMs calculation [33] between the upper excited $A\ ^1\Sigma^+\ (2(0^+))$ state and the three Ω components ($\Omega = 0^+, 0^-$, and 1) of the $X\ ^1\Sigma^+$ and $a\ ^3\Sigma^+$ states at the ground-state dissociation limit.

The electronic TDM operator $\boldsymbol{D}_{ij} = \langle i|e\Sigma_i r_i|j\rangle$ between an initial electronic state, |i⟩, and a final state, |j⟩, is associated with the transition between the two electronic states and electric dipole moment operator $\boldsymbol{\mu} = e\ \Sigma_i r_i$. Selection rules for electronic transitions are governed by $\Delta\Lambda = 0, \pm1$, $\Delta S = 0$, $\Delta\Omega = 0, \pm1$, $g \leftrightarrow u$, $\Sigma^+ \leftrightarrow \Sigma^+$, $\Sigma^- \leftrightarrow \Sigma^-$, and $\Sigma^- \leftarrow/\rightarrow \Sigma^+$ [37].

The UM− spectra in section 3.3.1 did not show transitions to the $A\ ^1\Sigma^+$ or $3\ ^1\Sigma^+$ states. However, the UM+ spectra include several transitions to vibrational levels of the $A\ ^1\Sigma^+$ state with term energies between $15\,116$ and $16\,225\,\mathrm{cm}^{-1}$ above the minimum of the ground $X\ ^1\Sigma^+$ state. If a $3(0^+)$ level is formed by PA, it can decay only to the $\Omega = 1$ component of the $a\ ^3\Sigma^+$ state or to the $X\ ^1\Sigma^+$ state ($\Omega = 0^+$), because transitions between the $1(0^-)$ and $3(0^+)$ states are forbidden. The dominant transition from our specific level of the $3(0^+)$ state to the $X\ ^1\Sigma^+$ state is to $v'' = 89$, and the dominant transition to the $a\ ^3\Sigma^+$ state is to $v_a = 21$.

Kotochigova *et al* calculated TDMs (figures 6 and 7 of [33], showing the TDM between the $1(1)$ and $2(0^+)$ states, and between the $1(0^+)$ and $2(0^+)$ states, respectively), which are large between the $X\ ^1\Sigma^+$ and $A\ ^1\Sigma^+$ state in the region of FC overlap between $v'' = 89$ and the $A\ ^1\Sigma^+$ state vibrational levels in the energy region between $15\,116$ and $16\,225\,\mathrm{cm}^{-1}$ above the minimum of the ground $X\ ^1\Sigma^+$ state. However, the TDM between $v_a = 21$ of $a\ ^3\Sigma^+$ ($\Omega = 1$) state and these vibrational levels of the $A\ ^1\Sigma^+$ state is quite small.

If instead a $3(0^-)$ level is formed by PA, it can only decay to the $a\ ^3\Sigma^+$ state ($\Omega = 1$ and 0^- components) because transitions between the $3(0^-)$ and $1(0^+)$ states are forbidden. As mentioned previously, most of the $a\ ^3\Sigma^+$ state population is in the level $v_a = 21$. Electric dipole transitions are not allowed between the $\Omega = 0^-$ component of the $a\ ^3\Sigma^+$ state and the $A\ ^1\Sigma^+$ ($\Omega = 0^+$) state. Also, the TDM for the $\Omega = 1$ component is quite small in the region of FC overlap between $v_a = 21$ and the energetically accessible levels of the $A\ ^1\Sigma^+$ state. We may apply the same argument as used for the $A^1\Sigma^+$ state to explain the absence of the $3\ ^1\Sigma^+$ state in the UM− spectra. Unfortunately there are no TDM calculations between the $4(0^+)$ and $1(1)$ states, or between the $4(0^+)$ and $1(0^+)$ states. If those calculations were available, it would help us to clarify the reason why the $3\ ^1\Sigma^+$ state is also missing in the UM− spectra, but present in the UM+ spectra. Note that Beuc *et al* [27] calculated TDMs for the $A\ ^1\Sigma^+ -$ $X\ ^1\Sigma^+$ and $3\ ^1\Sigma^+ - X\ ^1\Sigma^+$ transitions without considering the spin–orbit coupling.

3.5 Prospects and applications to other alkali metal diatomic molecules

The MB × UM multiplication method does not need detailed assignments of the intermediate states. Thus it is more helpful to investigate complex spectra of poorly studied molecules for population transfer and to analyze complicated spectra that cannot be analyzed by using the MB or UM spectra only. The MB × UM method should be very useful for other alkali heteronuclear diatomic molecules such as LiNa [39], LiK [40], LiCs [41], NaRb [42], NaCs [43], and RbCs [44] for which PA or MA has been carried out.

Let us consider another molecule, RbCs, as an example of using this method. In the RbCs case, excited spectra from both MB and UM experiments are available in the visible laser frequency region. Although PA and incoherent transfer to the ground state were carried out by Kerman *et al* [44] and Sage *et al* [45], there is still a need to find efficient transfer routes to the lowest level of the ground state by SRT. This reflects the fact that optimal transfer routes to the lowest level of the ground state are still not known. Debatin *et al* [46] used dark-state spectroscopy to investigate the intermediate $A\ ^1\Sigma^+ \sim b\ ^3\Pi$ states, mixed by spin–orbit interaction for transferring RbCs Feshbach molecules to the lowest vibrational level of the $X\ ^1\Sigma^+$ state. Also a recent PA experiment [47] has been carried out with high resolution in the region between $15\,363\ \mathrm{cm}^{-1}$ and $15\,735\ \mathrm{cm}^{-1}$ to investigate the $2\ ^3\Pi_0{}^+$ state for transferring to the lowest level of the ground state. MB experiments between $15\,475\ \mathrm{cm}^{-1}$ and $16\,005\ \mathrm{cm}^{-1}$ [48], and between $19\,550\ \mathrm{cm}^{-1}$ and $20\,575\ \mathrm{cm}^{-1}$ [49] from the lowest energy of the $X\ ^1\Sigma^+$ state for RbCs have been carried out although the intermediate region between $15\,980\ \mathrm{cm}^{-1}$ and $19\,525\ \mathrm{cm}^{-1}$ is not yet available. We expect that it may be possible to combine the MB and UM spectra in this energy region after deconvolution of the rotational structure and hyperfine structure. Then our MB × UM product spectra can be used to identify appropriate SRT routes.

Our combination method is very useful in assigning vibrational levels in perturbed spectral regions. We successfully employed this method to assign vibrational levels of the perturbed electronic states as discussed in chapter 4 in the case of KRb. There are existing regions which have not been analyzed so far. As an example, the MB spectra are simpler than those obtained by other experimental methods, but they could not be assigned to the vibrational levels of the specific electronic states because vibrational progressions for the electronic states in that region are not regular, due to large perturbations among the electronic states. However, if we use our combination method, such a severely perturbed region should be identified and optimal transfer routes to the lowest vibrational level of the $X\ ^1\Sigma^+$ state can be chosen from the strong singlet-triplet mixing with the intermediate states.

This method can also be applied to homonuclear alkali metal diatomic molecules with inversion symmetry. In this case, *gerade* excited states formed from PA decay to the *ungerade* $a \, ^3\Sigma_u^+$ state according to selection rules, while *ungerade* excited states decay to the *gerade* ground $X \, ^1\Sigma_g^+$ state. These two $X \, ^1\Sigma_g^+$ and $a \, ^3\Sigma_u^+$ states with different inversion symmetries cannot have common upper states, in contrast to heteronuclear diatomic molecules assuming electric-dipole-allowed transitions, and cannot be coupled by an SRT. However, it is possible to have hyperfine mixing between the $X \, ^1\Sigma_g^+$ and $a \, ^3\Sigma_u^+$ states formed extremely close to dissociation as in MA.

Thus the high-lying rovibrational levels of the $X \, ^1\Sigma_g^+$ state near its dissociation limit and intermediate electronic states with *ungerade* symmetry are needed to transfer the population to the lowest level of the ground state because the lowest level for population transfer is the singlet and *gerade* $X \, ^1\Sigma_g^+$ state. Thus high vibrational levels of only the $X \, ^1\Sigma_g^+$ state can be transferred by stimulated Raman transfer (SRT) to the lowest level of the ground state; high vibrational levels of the $a \, ^3\Sigma_u^+$ state cannot be used.

References

[1] Walter J M and Barratt S 1928 The existence of intermetallic compounds in the vapor state. The spectra of the alkali metals, and of their alloys with each other *Roy. Soc. Proc.* A **105** 257

[2] Amiot C 2000 The KRb (2) $^3\Sigma^+$ Electronic state *J. Mol. Spectrosc.* **203** 126

[3] Pashov A, Docenko O, Tamanis M, Ferber R, Knöeckel H and Tiemann E 2007 Coupling of the $X \, ^1\Sigma^+$ and $a \, ^3\Sigma^+$ states of KRb *Phys. Rev.* A **76** 022511

[4] Amiot C and Verges J 2000 The KRb ground electronic state potential up to 10 Å *J. Chem. Phys.* **112** 7068

[5] Ross A J, Effantin C, Crozet P and Boursey E 1990 The ground state of KRb from laser-induced fluorescence *J. Phys. B: At. Mol. Opt. Phys.* **23** L247

[6] Amiot C, Verges J, d'Incan J and Effantin C 2000 The $3 \, ^1\Pi - 3 \, ^1\Sigma^+$ system of KRb *Chem. Phys. Lett.* **315** 55

[7] Okada N, Kasahara S, Ebi T, Baba M and Katô H 1996 Optical–optical double resonance polarization spectroscopy of the $B \, ^1\Pi$ state of ^{39}K^{85}Rb *J. Chem. Phys.* **105** 3458

[8] Kasahara S, Fujiwara C, Okada N, Katô H and Baba M 1999 Doppler-free optical-optical double resonance polarization spectroscopy of the KRb $1 \, ^1\Pi$ and $2 \, ^1\Pi$ states *J. Chem. Phys.* **111** 8857

[9] Amiot C, Verges J, Effantin C and d'Incan J 2000 The KRb $2 \, ^1\Pi$ electronic state *Chem. Phys. Lett.* **321** 21

[10] Lee Y, Yun C, Yoon Y, Kim T and Kim B 2001 The 530 nm system of KRb observed in a pulsed molecular beam: New electric quadrupole transitions ($1 \, ^1\Delta - X \, ^1\Sigma^+$) *J. Chem. Phys.* **115** 7413

[11] Lee Y, Yoon Y, Kim B, Li L and Lee S 2004 Observation of the $3 \, ^3\Sigma^+ - X \, ^1\Sigma^+$ transition of KRb by resonance enhanced two-photon ionization in a pulsed molecular beam: Hyperfine structures of ^{39}K^{85}Rb and ^{39}K^{87}Rb isotopomers *J. Chem. Phys.* **120** 6551

[12] Lee Y, Yoon Y, Muhammad A, Kim J T, Lee S and Kim B 2010 The 480 nm system of KRb: The $1 \, ^3\Delta_1$, $4 \, ^1\Sigma^+$ and $5 \, ^1\Sigma^+$ states *J. Phys. Chem.* A **114** 7742

[13] Lee Y, Yoon Y, Kim J T, Lee S and Kim B 2011 Unravelling complex spectra of a simple molecule: REMPI study of 420 nm system of KRb *Chem. Phys. Chem.* **12** 2018

[14] Wang D, Qi J, Stone M F, Nikolayeva O, Hattaway B, Gensemer S D, Wang H, Zemke W T, Gould P L, Eyler E E and Stwalley W C 2004 The photoassociative spectroscopy, photo-associative molecule formation, and trapping of ultracold $^{39}K^{85}Rb$ *Eur. Phys. J.* D **31** 165

[15] Wang D, Qi J, Stone M F, Nikolayeva O, Wang H, Hattaway B, Gensemer S D, Gould P L, Eyler E E and Stwalley W C 2004 Photoassociative production and trapping of ultracold KRb molecules *Phys. Rev. Lett.* **93** 243005

[16] Wang D, Eyler E E, Gould P L and Stwalley W C 2006 Spectra of ultracold KRb molecules in near-dissociation vibrational levels *J. Phys. B: At. Mol. Opt. Phys.* **39** S849

[17] Wang D, Kim J T, Ashbaugh C, Eyler E E, Gould P L and Stwalley W C 2007 Rotationally resolved depletion spectroscopy of ultracold KRb molecules *Phys. Rev.* A **75** 032511

[18] Kim J T, Wang D, Eyler E E, Gould P L and Stwalley W C 2009 Spectroscopy of $^{39}K^{85}Rb$ triplet excited states using ultracold $a\,^3\Sigma^+$ state molecules formed by photoassociation *New J. Phys.* **11** 055020

[19] Banerjee J, Rahmlow D, Carollo R, Bellos M, Eyler E E, Gould P L and Stwalley W C 2013 Spectroscopy and applications of the 3 $^3\Sigma^+$ electronic state of $^{39}K^{85}Rb$ *J. Chem. Phys.* **139** 174316

[20] Banerjee J, Rahmlow D, Carollo R, Bellos M, Eyler E E, Gould P L and Stwalley W C 2013 Spectroscopy of the double minimum 3 $^3\Pi_\Omega$ electronic state of $^{39}K^{85}Rb$ *J. Chem. Phys.* **138** 164302

[21] Aikawa K, Akamatsu D, Hayashi M, Kobayashi J, Ueda M and Inouye S 2011 Predicting and verifying transition strengths from weakly bound molecules *Phys. Rev.* A **83** 042706

[22] Kim J T, Lee Y, Kim B, Wang D, Stwalley W C, Gould P L and Eyler E E 2011 Spec-troscopic analysis of the coupled 1 $^1\Pi$, 2 $^3\Sigma^+(\Omega = 0^-, 1)$, and $b\,^3\Pi$ $(\Omega = 0^\pm, 1, 2)$ states of the KRb molecule using both ultracold molecule and molecular beam experiments *Phys. Chem. Chem. Phys.* **13** 18755

[23] Kim J T, Lee Y, Kim B, Wang D, Gould P, Eyler E and Stwalley W 2012 Spectroscopic investigation of the A and 3 $^1\Sigma^+$ states of $^{39}K^{85}Rb$ *J. Chem. Phys.* **137** 244301

[24] Kim J T, Lee Y, Kim B, Wang D, Stwalley W C, Gould P L and Eyler E E 2011 Spec-troscopic prescription for optimal stimulated Raman transfer of ultracold heteronuclear molecules to the lowest rovibronic level *Phys. Rev.* A **84** 062511

[25] Kim J T 2013 Population transfer routes to the lowest vibrational level of ultracold $^{39}K^{85}Rb$ *J. Kor. Phys. Soc.* **63** 933

[26] Stwalley W C, Banerjee J, Bellos M, Carollo R, Recore M and Mastroianni M 2010 Resonant coupling in the heteronuclear alkali dimers for direct photoassociative formation of $X(0,0)$ ultracold molecules *J. Phys. Chem.* A **114** 81

[27] Beuc R, Movre M, Ban T, Pichler G, Aymar M, Dulieu O and Ernst W E 2006 Predictions for the observation of KRb spectra under cold conditions *J. Phys. B: At. Mol. Opt. Phys.* **39** S1191

[28] Tiemann E Private communication

[29] Kim J T, Stolyarov A V and Stwalley W C 2014 Spin-Orbit Coupling in the $2^3\Pi$ state of $^{39}K^{85}Rb$ (to be submitted)

[30] Breford E J and Engelke F 1978 Laser-induced molecular fluorescence in supersonic nozzle beams: Applications to the NaK $D\,^1\Pi$ - $X\,^1\Sigma^+$ and $D\,^1\Pi$ - $a\,^3\Sigma^+$ system *Chem. Phys. Lett.* **53** 282

[31] Lefebvre-Brion H and Field R W 1986 *Perturbations in the Spectra of Diatomic Molecules* 2nd edn (New York: Academic)

[32] Banerjee J, Rahmlow D, Carollo R, Bellos M, Eyler E E, Gould P L and Stwalley W C 2012 Direct photoassociative formation of ultracold KRb molecules in the lowest vibrational levels of the electronic ground state *Phys. Rev.* A **86** 053428

[33] Kotochigova S, Julienne P S and Tiesinga E 2003 *Ab initio* calculation of the KRb dipole moments *Phys. Rev.* A **68** 022501

[34] Supplementary material at http://dx.doi.org/10.1063/1.4771661

[35] Rousseau S, Allouche A R and Aubert-Frécon M 2000 Theoretical study of the electronic structure of the KRb molecule *J. Mol. Spectrosc.* **203** 235

[36] Supplementary material of reference [22].

[37] Herzberg G 1989 *Molecular Spectra and Molecular Structure I. Spectra of Diatomic Molecules* 2nd edn (Florida: R E Krieger)

[38] Mulliken R S 1960 The interaction of differently excited like atoms at large distances *Phys. Rev.* **120** 1647

[39] Heo M-S, Wang T T, Christensen C A, Rvachov T M, Cotta D A, Choi J-H, Lee Y-R and Ketterle W 2012 Formation of ultracold fermionic NaLi Feshbach molecules *Phys. Rev.* A **86** 021602(R)

[40] Ridinger A, Chaudhuri S, Salez T, Fernandes D R, Bouloufa N, Dulieu O, Salomon C and Chevy F 2011 Saturation in heteronuclear photoassociation of ^{6}Li^{7}Li *Europhysics Lett.* **96** 33001

[41] Deiglmayr J, Grochola A, Repp M, Mörtlbauer K, Glück C, Lange J, Dulieu O, Wester R and Weidemüller M 2008 Photoassociative creation of ultracold heteronuclear ^{6}Li^{40}K* molecules *Phys. Rev. Lett.* **101** 133004

[42] Xiong D, Li X, Wang F and Wang D 2013 Observation of Feshbach resonances between ultracold Na and Rb atoms *Phys. Rev.* A **87** 050702(R)

[43] Haimberger C, Kleinert J, Bhattacharya M and Bigelow N P 2004 Formation and detection of ultracold ground-state polar molecules *Phys. Rev.* A **70** 021402

[44] Kerman A J, Sage J M, Sainis S, Bergeman T and DeMille D 2004 Production of ultracold, polar RbCs* molecules via photoassociation *Phys. Rev. Lett.* **92** 033004

[45] Sage J M, Sainis S, Bergeman T and DeMille D 2005 Optical production of ultracold polar molecules *Phys. Rev. Lett.* **94** 203001

[46] Debatin M, Takekoshi T, Rameshan R, Reichsöllner L, Ferlaino F, Grimm R, Vexiau R, Bouloufa N, Dulieu O and Nägerl H-C 2011 Molecular spectroscopy for ground-state transfer of ultracold RbCs molecules *Phys. Chem. Chem. Phys.* **13** 18926

[47] Bruzewicz C D, Gustavsson M, Shimasaki T and DeMille D 2013 Continuous formation of vibronic ground state RbCs molecules via photoassociation *New J. Phys.* **16** 023018

[48] Lee Y, Yoon Y, Lee S, Kim J T and Kim B 2008 Parallel and coupled perpendicular transitions of RbCs 640 nm system: mass-resolved resonance enhanced two-photon ionization in a cold molecular beam *J. Phys. Chem.* A **112** 7214

[49] Lee Y, Yoon Y, Lee S and Kim B 2009 500 nm system of RbCs: Assignments and intensity anomalies *J. Phys. Chem.* A **113** 12187

IOP Concise Physics

Analysis of the Alkali Metal Diatomic Spectra
Using molecular beams and ultracold molecules
Jin-Tae Kim, Bongsoo Kim and William C Stwalley

Chapter 4

Multiplication spectroscopy for Raman transfer route determination

Molecules for laser excitation can be prepared in a heat-pipe oven, MB, ultracold molecule chamber, etc. The molecules in the heat-pipe oven are in a thermal Boltzmann population distribution, where many vibrational levels of the ground state are populated. It is not easy to prepare molecules only in low rotational levels. A supersonic MB beam method prepares internally cold molecules although translational velocities are still too high to trap the molecules. The rotational level of maximum population, J''_{max}, of the molecules is much lower than those in a heat-pipe oven.

UM molecules can be prepared from ultracold atoms with low kinetic energy so that molecules with rotational levels can be formed. Thus MB and UM molecules can be appropriate candidates for investigating low rotational levels. However, the UM molecules are formed at long range instead of near the equilibrium distance. For our purpose we need an excitation probability to rovibrational levels of excited states from the lowest ground level. Thus the MB experiment is the best way to obtain that kind of information.

Population transfer to the lowest ground state level from Feshbach and photo-associated molecules can be achieved. Feshbach molecules are primarily triplet in character and very near the dissociation limit of the ground state. PA provides triplet and singlet states from decay from PA levels depending on the symmetry of the PA levels.

Population in high vibrational levels of the $a\,^3\Sigma^+$ and $X\,^1\Sigma^+$ states formed by PA or Feshbach resonances have been transferred to the lowest level of the ground state through intermediate singlet and triplet states. However, for an $a\,^3\Sigma^+$ initial state an intermediate triplet electronic state must be mixed with an intermediate singlet electronic state due to spin selection rules, because the ground state is a pure singlet state. Also the symmetries (+, −) of PA levels may be considered because a photo-associated level with − symmetry is not allowed to decay to the states with + symmetry including the $X\,^1\Sigma^+$ state. However, a PA level with + symmetry is

4-1

allowed to decay to the states with + symmetry so that it can be allowed to decay to the $X\,^1\Sigma^+$ and $a\,^3\Sigma^+$ states.

Vibrational levels with strong multiplication intensities in the combined MB $\times$ UM spectra in the same energy range were searched to find appropriate strong transfer routes. This new method would be quite effective for finding the optimal SR pathways to transfer UM formed by PA or MA to the lowest rovibronic level. In our case MB and UM spectra vibrational levels include only low Js because of internal cooling of MB and UMs formed from ultracold atoms. The process of SR transfer is difficult to implement for an arbitrary combination of different alkali atoms, since the spectroscopy is complex, with many mutually perturbing electronic states, and it may be difficult to assign in the most favorable spectral regions for transfer. Fortunately, high quality PECs are now available for all ten heteronuclear species to estimate the most favorable regions [1]. Further investigations are needed to systematically analyze the formation of ultracold KRb molecules in the lowest rovibrational level ($v'' = 0$ and $J'' = 0$) through a broader range of intermediate electronic states. For these purposes, strong and efficient intermediate transition lines for the population transfer should be investigated.

4.1 Determination of the Raman transfer route using the 4 $^1\Sigma^+$ state

In this section we use the singlet 4 $^1\Sigma^+$ state instead of a singlet–triplet mixed state in determining the optimal intermediate transfer route to the lowest energy level of the ground state of KRb using the multiplication spectra between MB and UM spectra. There are two choices for reaching the lowest energy level of the ground state through inner and outer turning points of the 4 $^1\Sigma^+$ state. However, prediction of the unusual potential well shape of the 4 $^1\Sigma^+$ state is not easy, although theoretical calculation methods for the potential wells are now well developed. So it is best to use experimentally observed term energies with strong transition probabilities instead of calculated results for population transfers. Accordingly, we use experimentally observed spectroscopic lines of the 4 $^1\Sigma^+$ state obtained from the MB [2] and UM [3] methods to determine the transfer routes to the lowest level of the ground state. The state has a broad potential well that differs from other states with narrower potential wells near the equilibrium point of the ground state potential. Therefore, it is good to use this state to transfer population from larger to smaller internuclear distances.

There was an investigation into KCs [4] using the intermediate 4 $^1\Sigma^+$ state with a shelf potential well from weakly bound rovibrational levels of both the singlet $X\,^1\Sigma^+$ and the triplet $a\,^3\Sigma^+$ state to produce the lowest rovibrational level. They used experimental and theoretical absorption and emission Einstein coefficients between the 4 $^1\Sigma^+$ and $X\,^1\Sigma^+$ and $a\,^3\Sigma^+$ states.

The 1 $^3\Delta$, 4 $^1\Sigma^+$, and 5 $^1\Sigma^+$ states of KRb in the energy range between 19750 and 21600 cm^{-1} were observed from the $X\,^1\Sigma^+$ state using RE2PI [2]. However, the 4 $^3\Sigma^+$ and 3 $^3\Pi$ triplet states in the same energy region as the 1 $^3\Delta$, 4 $^1\Sigma^+$, and 5 $^1\Sigma^+$ states were not observed in the MB experiment, although the FCFs between the $v'' = 0$ level of the $X\,^1\Sigma^+$ state and vibrational levels of 4 $^3\Sigma^+$ and 3 $^3\Pi$ states are comparable or larger than those to vibrational levels of the $4\ ^1\Sigma^+$ and $5\ ^1\Sigma^+$ singlet states. These triplet states were not observed as they were spin-forbidden from the ground singlet state and because there is no significant mixing between those triplet states and nearby singlet states. However, Wang $et\ al$ [3] investigated the $4\ ^3\Sigma^+$ and $4\ ^1\Sigma^+$ states excited from highly excited long-range vibrational levels of the $X\ ^1\Sigma^+$ and $a\ ^3\Sigma^+$ states, in contrast to $v'' = 0$ and $v'' = 1$ in the MB experiment of Lee $et\ al$ [2].

The lowest observed vibrational quantum number of the $4\ ^1\Sigma^+$ state from the MB spectra in [2] was assigned to $v'_{min} = 31$, but the corresponding vibrational quantum number determined from the UM spectra in [3] is $v'_{min} = 27$. Thus, there is a difference of four vibrational quanta in vibrational quantum number assignments between two different (MB and UM) experiments. Although there are such differences in the vibrational quantum number assignments between MB and UM spectra, there are quite good matches in the vibrational energies between the MB and UM spectra.

We add the eigenvalues of the $4\ ^1\Sigma^+$ state (v', $J' = 0$) obtained from the theoretical potential curve [5] referenced to the first dissociation limit of the ground state to precisely determine the experimental dissociation energy of the ground state [6]. As shown in table 4.1, theoretical term energies match well with experimental term energies in [2] and [3] when we chose the vibrational level $v'_{min} = 27$ from [4] and $v'_{min} = 31$ from [2] as $v'_{min} = 29$. Nice matches are found for the experimental and theoretical term values for $v' = 37\text{–}44$ ($\sim 0.5\ cm^{-1}$) although there are mismatches of several cm^{-1} for other vibrational levels, as shown in table 4.1.

Experimental MB spectra from [2], UM spectra from [3], and the multiplication spectra between the MB and UM spectra are shown in figures 4.1(a), (b), and (c), respectively. We use the theoretical vibrational quantum numbering ($v'_{min} = 29$) discussed above and the spectral intensity for each vibrational quantum number from MB and UM experiments to find the optimal vibrational quantum number for the population transfer. The observed MB and UM spectra have quite broad distributions up to higher vibrational quantum numbers. From figure 4.1(c) the maximum transfer rate (which has the maximum multiplication value between the MB and UM) is at $v' = 38$. The multiplication feature for the transfer has a broad vibrational distribution due to the broad potential near the bottom of the state compared to those investigated previously in [8] because that is due to perturbations for specific vibrational levels. Using the multiplicative product spectra method, optimal intermediate vibrational levels through the $4\ ^1\Sigma^+$ state of $^{39}K^{85}Rb$ for formation of the lowest vibrational level of $X\ ^1\Sigma^+$ state from high vibrational levels of the ground state formed by decay from photoassociated UMs have been determined.

4.2 Determination of the Raman transfer route using singlet and triplet mixed states

Figure 4.2(c) shows the multiplicative product between the UM spectra in figure 4.2(b) and MB spectra in figure 4.2(a) between 15410 and 15610 cm^{-1} with their assignments from spectral analysis described in sections 3.3.1 and 3.3.2. MB spectra in figure 4.2(a) show that the singlet $1\ ^1\Pi$ state excited from the singlet $X\ ^1\Sigma^+$ state with populations in

Table 4.1. $T_{v'}$ values of the vibrational level energies of the 4 $^1\Sigma^+$ state from references [2] and [3], and theoretical potentials. We add the dissociation energy of the X $^1\Sigma^+$ state, $4217.822\,\text{cm}^{-1}$ [6], to the corresponding values reported in reference [3] in order to convert them to the $T_{v'}$ values. For the theory column, eigenvalues of the 4 $^1\Sigma^+$ states (v', $J' = 0$) from the theoretical potential curves in reference [10] are added to the dissociation energy [6] of the X $^1\Sigma^+$ state to obtain theoretical $T_{v'}$ values of the 4 $^1\Sigma^+$ state.

Reference [2]		Reference [3]		Theory	
v'	$T_{v'}$ (cm^{-1})	v'	$T_{v'}$ (cm^{-1})	v'	$T_{v'}$ (cm^{-1})
31	20473.8	27	20473.5	29	20477.9
32	20490.8	28	20490.7	30	20494.6
33	20508.1	29	20507.9	31	20511.5
34	20525.6	30	20525.3	32	20528.5
35	20543.1	31	20543.0	33	20545.8
36	20560.9	32	20561.2	34	20563.3
37	20579.0	33	20579.2	35	20581.0
38	20597.0	34	20597.2	36	20598.8
39	20615.3	35	20615.6	37	20616.8
40	20633.8	36	20634.0	38	20635.0
41	20652.4	37	20652.8	39	20653.3
42	20671.0	38	20671.6	40	20671.8
43	20690.0	39	20690.4	41	20690.5
44	20709.0	40	20709.2	42	20709.3
45	20728.3	41	20728.5	43	20728.3
46	20747.5	42	20747.6	44	20747.4
47	20766.8	43	20767.2	45	20766.6
48	20786.2	44	20786.4	46	20785.9
49	20806.1	45	20806.7	47	20805.4
50	20825.7	46	20826.4	48	20825.0
51	20845.6	47	20846.2	49	20844.7
52	20865.7	48	20866.3	50	20864.5
53	20885.6	49	20886.2	51	20884.4
54	20905.9	50	20906.3	52	20904.5
55	20926.1	51	20926.8	53	20924.6
56	20946.3	52	20947.4	54	20944.8
57	20966.9	53	20967.9	55	20965.1
58	20987.3	54	20988.4	56	20985.6
59	21008.0	55	21009.1	57	21006.0
60	21029.1	56	21029.2	58	21026.6
61	21049.5	57	21050.5	59	21047.3
62	21070.4	58	21071.4	60	21068.0

$v'' = 0$ and $v'' = 1$ has the strongest signals although other singlet states, such as the A $^1\Sigma^+$ state with small FCFs and 3 $^1\Sigma^+$ state with weak TDMs, have very weak signals. The triplet states such as the 2 $^3\Sigma^+$ and b $^3\Pi$ states have strong signals only for vibrational levels perturbed locally by nearby levels of the 1 $^1\Pi$ state, but other triplet levels show weak spectra due to weak mixing with singlet states.

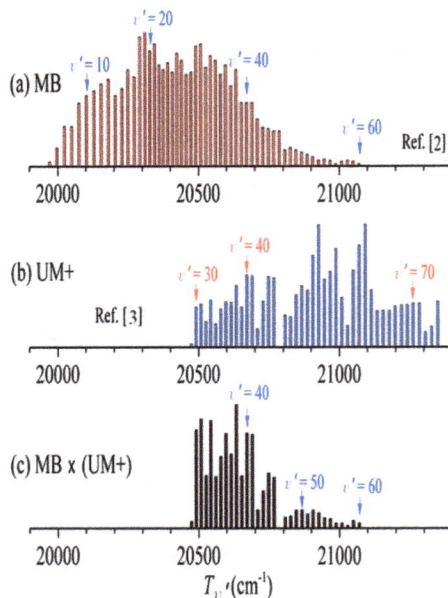

Figure 4.1. (*a*) Experimental MB spectra from reference [2] and (*b*) UM+ spectra from reference [3], and (*c*) the multiplication spectra between the MB and UM+ spectra. We expect that the maximum transfer rate which has maximum multiplication value between the MB and UM is available at $v' = 38$, which is near the maximum expected FCF at $v' = 40$. Reproduced from [7].

However, in the UM− spectra the initial state for excitation is the triplet $a\ ^3\Sigma^+$ state with population in the $v_a = 20$ and $v_a = 21$ levels. Thus in UM− excitation spectra, transitions to the excited triplet states are favorable compared to transitions to singlet states. Consequently, we expect proper intermediate states for population transfer to be mixed states between singlet and triplet states and with strong signals in both UM− and MB spectra.

Considering the spectroscopic selection rules, we expect that the triplet–triplet transitions will appear as the main transitions in the RE2PI spectra from UM and the singlet–singlet transitions in the RE2PI spectra from MB. Because of singlet–triplet mixing, however, identical states can be reached by both transitions. The main features of the RE2PI UM spectra from the $a\ ^3\Sigma^+$ state to the $2\ ^3\Sigma^+$ and $b\ ^3\Pi$ states and those of RE2PI MB spectra from the $v'' = 0$ level of the $X\ ^1\Sigma^+$ state to $v' = 0$–14 levels of the $1\ ^1\Pi$ state are readily identified by considering the corresponding FCFs calculated using the *ab initio* PECs and the electric dipole transition moments reported by Beuc *et al* [9].

We found that the optimal SR pathways in the region 15409–15600 cm^{-1} are established when the intermediate levels located at 15452.99, 15459.55, 15511.90, and 15520.21 cm^{-1} from the minimum of the ground $X\ ^1\Sigma^+$ state potential are employed, as marked in figure 4.2(*c*) by α, β, γ, and δ, respectively. α and δ are assigned to the transitions to the vibrational levels of the $2\ ^3\Sigma^+(\Omega = 1)$ state. β and γ are transitions to vibrational $v' = 7$ and 8 levels of the $1\ ^1\Pi$ state, respectively.

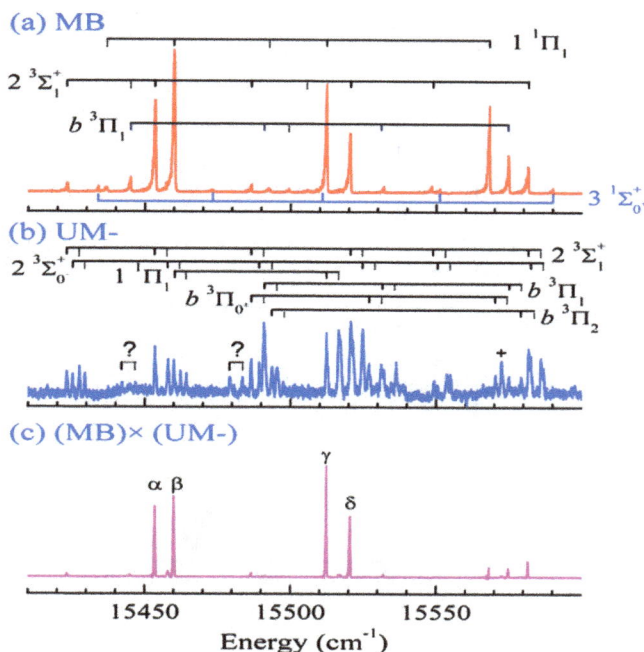

Figure 4.2. (*a*) MB, (*b*) UM−, and (*c*) product between MB and UM− spectra in the region 15410–15610 cm^{-1}. Assignments of the excited states such as $1\ ^1\Pi_1$, $b\ ^3\Pi_{0\pm}$, $b\ ^3\Pi_1$, $b\ ^3\Pi_2$, $2\ ^3\Sigma_1^+$, and $2\ ^3\Sigma_0^{-+}$ have been made. The pairs of lines marked with symbol (?) are not assigned and the single line marked with symbol (+) is also unassigned. The very weak lines of the $3\ ^1\Sigma^+$ state have recently been assigned [8]. The $X\ ^1\Sigma^+$ state molecules probed in the MB experiment are primarily in $v'' = 0$ and 1, split by 75.38 cm^{-1}, which are labeled by vertical solid and dotted lines in the MB spectra, respectively. The a state molecules probed in the UM− experiment are mainly in $v_a = 20$ and 21, split by 4.23 cm^{-1}, which are labeled by vertical dotted and solid lines in the UM− spectra, respectively. The energy shown is the energy with respect to the minimum of the ground state potential, so the UM− spectrum has been shifted by the energy of the $a\ ^3\Sigma^+$ state $v_a = 21$ level. The strongest products ((MB)×(UM−)), indicated by α, β, γ, and δ, are the optimal pathways for SRT, as discussed in the text.

γ would provide the most efficient population transfer route to the vibrational $v'' = 0$ level of the $X\ ^1\Sigma^+$ state through the intermediate $v' = 8$ level of the $1\ ^1\Pi$ state from the vibrational $v = 21$ level of the $a\ ^3\Sigma^+$ state. All four intermediate levels have significant singlet and triplet character.

This most efficient intermediate vibrational $v' = 8$ level of the $1\ ^1\Pi$ state for population transfer to the lowest energy of the ground state can be seen to be perturbed from other states using vibrational spacing $\Delta G_{v'+1/2}$ plot of the $1\ ^1\Pi$ state. However, for SRT, the perturbing state must be a triplet to contribute triplet character to this singlet state. In order to transfer the population from triplet to singlet states via an intermediate state, the intermediate state should have mixed singlet-triplet character. Figure 4.3 (expanded from figure 3.3) shows plots of vibrational level spacing between adjacent vibrational levels ($\Delta G_{v'+1/2} = G(v'+1) - G(v')$) versus term value $T_{v'} = T_{e'} + G(v')$ for the $1\ ^1\Pi$, $2\ ^3\Sigma^+$ ($\Omega = 1$), and $b\ ^3\Pi$ ($\Omega = 1$) states perturbing one another by spin–orbit

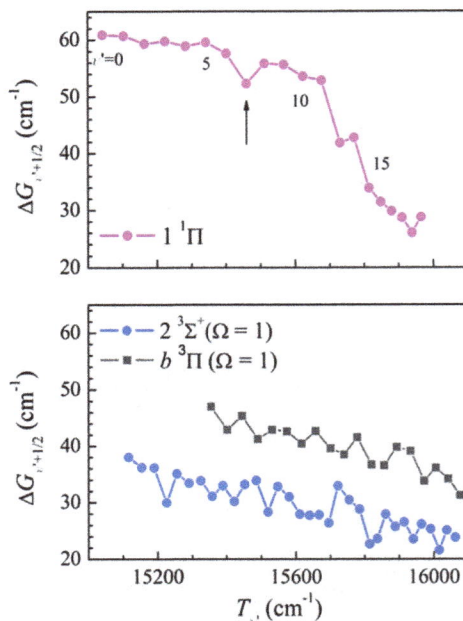

Figure 4.3. Vibrational intervals $\Delta G_{v'+1/2}$ as a function of $T_{v'}$ for (a) the 1 $^1\Pi$ state, and (b) the 2 $^3\Sigma^+(\Omega=1)$ and b $^3\Pi(\Omega=1)$ states. The perturbed $v'=8$ level in (a), corresponding to the arrow, proves to be the optimal SRT route because of its strongly mixed singlet-triplet character. Reproduced from [7].

interaction. Significant fluctuations in the $\Delta G_{v'+1/2}$ values indicate that the perturbations are strong, leading to a large singlet–triplet mixing. The $\Delta G_{v'+1/2}$ value for the $v'=7$ value of the 1 $^1\Pi$ state indicated by the arrow in figure 4.3(a) is also remarkably irregular. This irregularity of the $\Delta G_{v'+1/2}$ value for $v'=7$ is consistent with the appearance of strong extra bands at 15452.99 (α) and 15520.21 (δ) cm^{-1} in the MB RE2PI spectra. Comparable intensities of the vibrational $v'=7$ and $v'=8$ levels of the 1 $^1\Pi$ state from the $v''=0$ level of the $X^1\Sigma^+$ state and the extra bands suggest very strong mixing between the 1 $^1\Pi$ and 2 $^3\Sigma^+$ ($\Omega=1$) states. These analyses indicate that the singlet–triplet mixing among the $\Omega=1$ states (1 $^1\Pi$, 2 $^3\Sigma^+$ ($\Omega=1$), and b $^3\Pi$ ($\Omega=1$)) is particularly strong for the α, β, γ, and δ peaks of the product spectrum and this strong singlet–triplet mixing corresponds to efficient population transfer.

Multiplication spectra between the UM+ spectra using a PA level of the 3(0$^+$) state instead of the 3(0$^-$) state have also been obtained, as shown in figure 4.4. The four strong lines, marked as α', β', γ', and δ' in the multiplication spectra, show the same vibrational levels, marked as α, β, γ, and δ, respectively, in figure 4.2. The strongest β' line in the spectra corresponds to a favorable population transfer vibrational level of the singlet 1 $^1\Pi$ state. Initial vibrational wavefunctions in both cases are different as the TDM matrix elements between the initial and intermediate levels for the UM+ and UM− cases are. Thus, the intensities of the UM− spectrum shown in figure 4.2(b) are different from those of the UM+ spectrum shown in figure 4.4(b). The multiplication spectrum in the case of UM+ is different from the

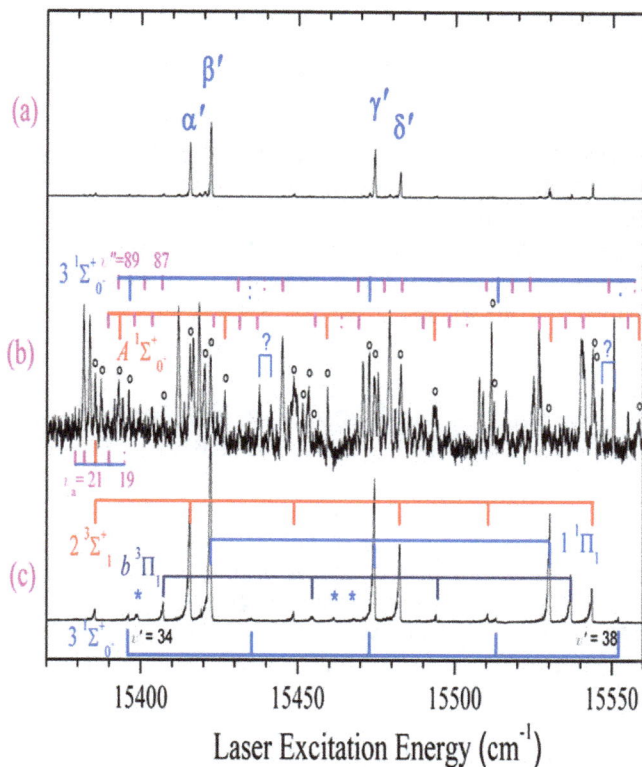

Figure 4.4. (*a*) Multiplication product spectra between MB and UM+ spectra in the region $15372 - 15562\,\text{cm}^{-1}$. (*b*) Assigned UM+ spectra of the $A\,{}^1\Sigma^+$, $3\,{}^1\Sigma^+$, $1\,{}^1\Pi$, $2\,{}^3\Sigma^+$, and $b\,{}^3\Pi$ levels excited from high vibrational levels of the $X\,{}^1\Sigma^+$ and $a\,{}^3\Sigma^+$ states formed by decay from a specific level of the $3(0^+)$ state (formed by PA) (*c*) Assigned MB spectra of the $3\,{}^1\Sigma^+$, $1\,{}^1\Pi$, $b\,{}^3\Pi$, and $2\,{}^3\Sigma^+$ levels excited from low vibrational levels ($v'' = 0$ and 1) of the $X\,{}^1\Sigma^+$ state. Hot bands from the $v'' = 1$ level of the $X\,{}^1\Sigma^+$ state vibrational levels are marked by the symbol (*). The strongest products (MB×UM), indicated by α', β', γ', and δ', are the optimal pathways for SRT, as discussed in the text.

multiplication spectrum in the case of UM−. If we extend the UM spectra using the $3(0^+)$ state to the higher energy region, we may find the appropriate transfer line of the $3\,{}^1\Sigma^+$ state found by Aikawa *et al* [10].

For several heteronuclear diatomic molecules, such as LiCs, RbCs, and KRb as described in this section, SRT has been achieved into the lowest vibrational level or the lowest rovibrational level. In the case of ${}^{40}\text{K}\,{}^{87}\text{Rb}$, Ospelkaus *et al* [11] obtained SRT efficiencies of over 90% from the initial Feshbach molecule with primarily triplet character to the rovibrational ground state in a high phase space density gas of KRb molecules. They used a lower vibrational level of the $2\,{}^3\Sigma^+$ state using a near infrared laser beam wavelength which is not in the strongly perturbed region studied. They formed molecules in the lowest hyperfine level of lowest rovibrational level transferred from the Feshbach molecules. Such a transfer route can be investigated by accessing the intermediate level using an infrared laser beam, but it is our expectation that the transition frequencies in figures 4.2 and 4.4 will be more efficient because of the high degree of significant mixing involved.

The first purely singlet transfer to the lowest rovibrational energy level was done by Aikawa *et al* [10]. They employed the STIRAP process in $^{41}K^{87}Rb$ for population transfer into the lowest rovibronic level using a pure singlet level of the $3\ ^1\Sigma^+$ state as the intermediate state from weakly bound levels of the $X\ ^1\Sigma^+$ state formed by PA. The energy region they used for population transfer was slightly above our investigated energy region. We expect our MB $\times$ UM product spectroscopy method would work equally well in identifying the optimal intermediate states for this singlet–singlet transfer if we investigated their energy region.

References

[1] Stwalley W C, Banerjee J, Bellos M, Carollo R, Recore M and Mastroianni M 2010 Resonant coupling in the heteronuclear alkali dimers for direct photoassociative formation of $X(0,0)$ ultracold molecules *J. Phys. Chem.* A **114** 81

[2] Lee Y, Yoon Y, Muhammad A, Kim J T, Lee S and Kim B 2010 The 480 nm system of KRb: The $1\ ^3\Delta_1$, $4\ ^1\Sigma^+$ and $5\ ^1\Sigma^+$ states *J. Phys. Chem.* A **114** 7742

[3] Wang D, Eyler E E, Gould P L and Stwalley W C 2006 Spectra of ultracold KRb molecules in near-dissociation vibrational levels *J. Phys. B: At. Mol. Opt. Phys.* **39** S849

[4] Klincare I, Nikolayeva O, Tamanis M, Ferber R, Pazyuk E A and Stolyarov A V 2012 Modeling of the $X\ ^1\Sigma^+$, $a\ ^3\Sigma^+ \rightarrow E(4)\ ^1\Sigma^+ \rightarrow X\ ^1\Sigma^+$ (v = 0, J = 0) optical cycle for ultracold KCs molecule production *Phys. Rev.* A **85** 062520

[5] Rousseau S, Allouche A R and Aubert-Frécon M 2000 Theoretical study of the electronic structure of the KRb molecule *J. Mol. Spectrosc.* **203** 235

[6] Wang D, Kim J T, Ashbaugh C, Eyler E E, Gould P L and Stwalley W C 2007 Rotationally resolved depletion spectroscopy of ultracold KRb molecules *Phys. Rev.* A **75** 032511

[7] Kim J T, Lee Y, Kim B, Wang D, Stwalley W C, Gould P L and Eyler E E 2011 Spectroscopic prescription for optimal stimulated Raman transfer of ultracold heteronuclear molecules to the lowest rovibronic level *Phys. Rev.* A **84** 062511

[8] Kim J T, Lee Y, Kim B, Wang D, Gould P, Eyler E and Stwalley W 2012 Spectroscopic investigation of the A and $3\ ^1\Sigma^+$ states of $^{39}K^{85}Rb$ *J. Chem. Phys.* **137** 244301

[9] Beuc R, Movre M, Ban T, Pichler G, Aymar M, Dulieu O and Ernst W E 2006 Predictions for the observation of KRb spectra under cold conditions *J. Phys. B: At. Mol. Opt. Phys.* **39** S1191

[10] Aikawa K, Akamatsu D, Hayashi M, Oasa K, Kobayashi J, Naidon P, Kishimoto T, Ueda M and Inouye S 2010 Coherent transfer of photoassociated molecules into the rovibrational ground state *Phys. Rev. Lett.* **105** 203001

[11] Ospelkaus S, Pe'er A, Ni K-K, Zirbel J j, Neyenhuis B, Kotochigova S, Julienne P S, Ye J and Jin D S 2008 Efficient state transfer in an ultracold dense gas of heteronuclear molecules *Nature Phys.* **4** 622

IOP Concise Physics

Analysis of the Alkali Metal Diatomic Spectra
Using molecular beams and ultracold molecules
Jin-Tae Kim, Bongsoo Kim and William C Stwalley

Chapter 5

Conclusions

We have reviewed previously developed supersonic MB machines and how those have been applied to molecular spectroscopy. In particular, pulsed supersonic MBs are very useful in investigating molecular spectra of alkali metal diatomic molecules because of low rovibrational temperatures of molecules. We have applied those in the case of the heteronuclear KRb molecule, and simple and strong spectra have been obtained.

Also, previous research into the generation of UMs has been reviewed. We have investigated excitation spectroscopy of UMs formed by decay from photoassociated KRb molecules. We have combined the supersonic MB spectroscopy with UM spectroscopy. From this combined $MB \times UM$ method we have investigated the following aspects in the case of KRb.

We introduce a new $MB \times UM$ product spectroscopy for identifying pathways suitable for STIRAP transfer of UMs to the lowest rovibronic level and demonstrate that in the case of KRb. This method does not require detailed spectroscopic assignments of the intermediate states used for transfer. High vibrational levels of complex perturbed spectra of the $A\ ^1\Sigma^+$, $3\ ^1\Sigma^+$, $1\ ^1\Pi$, $2\ ^3\Sigma_\Omega^+$, and $b\ ^3\Pi_\Omega$ states between 15 116 and 16 225 cm^{-1} above the minimum of the ground $X\ ^1\Sigma^+$ state of KRb have been observed and assigned, which cannot readily be analyzed by using the spectra from either MB or UM experiments alone.

The reversal of energy positions of Ω components in the $2\ ^3\Sigma^+$ state with respect to energy positions of the vibrational levels have been observed and assigned using the spectra from both the MB and the UM experiments and compared with theoretical calculations.

UMs formed by using the $3(0^-)$ PA level (called the UM– spectra) are not excited to the $A\ ^1\Sigma^+$ and $3\ ^1\Sigma^+$ states in the same energy region, but UMs formed by using the $3(0^+)$ PA level (called the UM+ spectra) are excited. The contrast between UM– and UM+ spectra of the $A\ ^1\Sigma^+$ and $3\ ^1\Sigma^+$ states has been explained by considering Hund's case (c) selection rules in conjunction with TDM calculations

doi:10.1088/978-1-6270-5678-6ch5

by Kotochigova *et al* [1] between the upper excited A $^1\Sigma^+$ $(2(0^+))$ state and the three Ω components at the ground-state dissociation limit.

Applications of this method to other alkali metal diatomic molecules where spectroscopic information is lacking or unassigned will also be very useful in finding efficient transfer routes to the lowest rovibrational level of the ground state of alkali metal polar UMs.

Reference

[1] Kotochigova S, Julienne P S and Tiesinga E 2003 *Ab initio* calculation of the KRb dipole moments *Phys. Rev.* A **68** 022501

www.ingramcontent.com/pod-product-compliance
Lightning Source LLC
Chambersburg PA
CBHW081554220326
41598CB00036B/6673